# The Flight of the

The story of Scottish pion
Captain Eric Starling F.R.Met.S., M.R.Ae.S.

Iain Hutchison



Published by
Kea Publishing

Printed by Essprint Ltd, Stornoway
and Nevisprint Ltd, Fort William

ISBN 0 9518958 0 X

*In Memory of*
*My Father*
*James Sanderson Hutchison*
*13 August 1912 — 22 September 1976*

# Contents

# Introduction

I first met Captain Eric Starling when I was compiling material for my book, The story of Loganair. I had undertaken that project because the growth of a fledgling Scottish airline that had been achieved initially from the pickings left over by the nationalised airline, which effectively left it to the country lanes of the sky serving the remotest of our communities, was a source of fascination to me.

However the names that Loganair's Islander aircraft carried also aroused my curiosity. Who had they been named after? One Loganair veteran doubted their relevance to a book on Loganair's development since none of them had been connected with the airline. But the choice of names honoured a heritage to which Loganair had fallen heir so I felt the topic was worth pursuing.

Of the seven names carried by the Loganair Islanders, five of the people honoured had passed to the great hangar in the sky. One, E. L. Gandar Dower, the legendary founder of Allied Airways, was in his nineties and in a declining state of health in Guernsey — he died at the age of 93 on 4th October 1987. The seventh name to be honoured by Loganair proved to be very much alive and kicking in the form of the ever youthful Eric Starling.

Eric, it transpired, was able to tell me not just about his colourful flying career but had a wealth of material, photographs and stories about many of the early Scottish airlines and their characters. His lively wit was accompanied by a depth of knowledge gleaned from four decades of flying everything from the diminutive BAC Drone that ran on a motor cycle engine to the four-engined Viscount airliner that was probably the ultimate in short-haul aircraft before their reign was eclipsed by the age of the jet.

Eric's greatest claims to fame are those associated with his pioneering days with Allied Airways. However he also enjoyed a colourful career during the Second World War. He claims to have had 'a pretty easy war', and this may be so in that his duties were primarily surveillance and training. But not only did he survive unscathed from a conflict which inflicted severe casualities on air crews, he undoubtedly helped numerous others emerge from those uncertain days without mishap to see more peaceful times.

In the post war period Eric was to spend twenty-five years flying with British European Airways from their Glasgow base. The last three years were spent flying for the Scottish Air Ambulance Service, a fitting conclusion to an illustrious career since it took him back to a type of flying very similar to that with which he had set out forty years earlier. When Eric retired he was B.E.A.'s most senior captain.

'Retired' of course is an inappropriate word to use since people like Eric never retire. While he may have a wealth of flying stories that he will launch into with little prompting, his ever active mind does not dwell too greatly on 'the good old days' and is always seeking new challenges. During one visit he mentioned that the Post Office were coming to see him about designs that he had proposed to them for an improved design of Post Box. His living room is a glittering array of computer terminals and computer programming has becoming an interest bordering on the fanatical.

While I was working on this book Eric would frequently wish me 'good luck' with a 'rather you than me'. But his contribution was very substantial since he not only waxed lyrical during our discussions but frequently my mail would bring bundles of print-outs of stories that had come to mind and he had instantly committed them to his word processor for me. 'The Flight of the Starling' really is Eric's story.

Iain Hutchison

# Foreword

*by Sir Peter Masefield*

CAPTAIN ERIC ALLEN STARLING, Master Pilot No. 166, a Founder Member of the British Airline Pilots Association, a Liveryman of the Guild of Air Pilots and Air Navigators, a great name in Scottish air transport and one of those very few who, worthily, can be termed 'a legend in his own lifetime', is the subject of this excellent book by Iain Hutchison. It brings to life a fascinating era in the story of British aviation.

Between 1934 and 1971, in the course of more than 12,500 flying hours, Eric Starling made history along many of the most important of Scotland's air routes. Moreover, he flew them in a remarkable range of aircraft — all the way from the little Shorts Scion of 1934 to the Vickers Viscount of the turbine era, the first turbine-powered aircraft to go into commercial airline service. In between them, in a six-year war-time interlude, Starling added to his log book a formidable list of military aircraft. They ranged from the Hawker Hurricane and Bristol Beaufighter to the Vickers-Armstrong Wellington and the Supermarine Walrus.

Starling's long and distinguished career in aviation began at Croydon Airport — the London Terminal Aerodrome of its day — when, at the age of just 20, in August 1931, he became an apprentice with the little Redwing Aircraft Limited. It came to an honourable conclusion, 40 years later, at Abbotsinch Airport, Glasgow, as B.E.A.'s most senior and respected airline captain. In all that time, no passenger received so much as a scratch in any aircraft under Eric Starling's command.

For six years, between 1949 and 1959, Eric and I shared together, in B.E.A. — in somewhat different ways — the building of a sound foundation for post-War British air transport. I well remember the first time we met. It was at Northolt Airport on a sunny morning in 1949 when Eric had flown an inaugural service there, from Renfrew, by way of Turnhouse, Edinburgh, accompanied by Bob McKean, B.E.A.'s General Manager, Scotland.

The reputation of the 38-year old Captain Starling was already something to be proud of — going back, as it did, to the somewhat hairy, pioneer over-water air services from Dyce Airport, Aberdeen, to Stromness in Orkney and on to Sumburgh in Shetland. Those were the days when early air transport in Scotland was being founded — with some rivalry — by two redoubtable, thoroughly independent and enterprising, characters — Eric Gandar Dower of Aberdeen Airways at Dyce and Edmund Fresson of Highland Airways at Longman Aerodrome, Inverness. I relate, with regret, that both Gandar Dower and Fresson were shabbily treated in the first post-War days of B.E.A. Their successful, but still struggling, air services were abruptly taken over by the nationalised airline whose Managing Director of the time — bred up in more affluent conditions — had little understanding of the hard struggles and personal involvement of the hardy Scottish pioneers. I have always been glad that — a little later on — I was able to do something to repair the shabby treatment and lack of consideration shown to those who had 'borne the heat and burden of the day' — difficult characters as they both were.

What was so remarkable about Eric Starling's career over the years was the way in which he made the transition — with skill and equanimity — from the early 'seat of the pants' flying, without blind flying instruments over difficult land and icy waters, to the quite different conditions of a management pilot (B.E.A.'s Flight Manager, Scotland, 1949-68) in command of fully equipped modern transport aircraft. In between, with another renowned Scottish pioneer pilot — David Barclay — Eric Starling did much to build up Scottish air ambulance services, often in atrocious weather. They were flown, at first, with DH84 Dragons, then with DH89A Rapides and, finally, with the much more suitable DH104 Herons.

This then is a remarkable story of courage, enterprise and ability, of extraordinary incidents and of many adventures in both war and peace — over Scotland, in Iceland, over the Western Desert and in India.

Uniquely, in his learning days, on a training night flight from Croydon to Lympne, Eric made an inadvertent but safe landing in a street in Calais, France, after unknowingly flying across the English Channel in the dark!

Iain Hutchison has given us a well researched and well told story in this excellent book. It pays a proper tribute to one of our classic pilots. It will, I know, be read with pleasure and appreciation.

I am delighted to have been invited to contribute this brief introduction to the story of a consummate pilot, a Flight Manager who was highly regarded by his 'troops' and, above all, a very nice man.

Peter Masefield
Reigate,
Surrey,
England

# Night Flight to Lympne

*'Are you the bloody fool who landed at night on a Calais street last week? You're as mad as I am. I'm going to build an airport and start air lines at Aberdeen. Are you willing to join me?'*

These are the words spoken by Eric Gander Dower, founder of Allied Airways, that brought Eric Starling to Scotland.

The year was 1933 and at the time Eric Starling was an apprentice with the Redwing Aircraft Company. There he had been licenced after the appropriate training on aircraft and their engines as a Ground Engineer. He also learned to fly, first of all qualifying for the Pilots 'A' Licence, the private pilots licence, and now he was taking the examinations for his Pilots 'B' Licence, the qualification for a Commercial Pilot.

One of the tests required in order to obtain the Pilots 'B' Licence involved a solo night flight. But it is best that Eric Starling tells his own story of the time he set out to undertake this examination.

'The first time in my life that I broke out of obscurity and made headlines in the newspapers was on the night of Tuesday 15th May 1933. At that time I was a pilot apprentice with the Redwing Aircraft Company who manufactured a small two-seater bi-plane, one of the first open two-seater aircraft to put the passenger or pupil beside the pilot.

'By the end of May 1933 I was at the end of my apprenticeship and taking the test for my 'B' licence. One of the tests was a night flight. This flight had to be done on a moonless night, either from Croydon to Lymnpe on the Kent coast, or the other way around. The Ministry made one carry a sealed barograph which indicated the height at which one had been flying, but more important, that you hadn't landed somewhere en route. The distance is about 55 miles, taking about 45 minutes at the Redwing's cruising speed of 75 miles per hour.

'Although I never heard of any one else having trouble, the insurance companies took a jaundiced view of these night tests and a policy to cover the tests cost more than the hire of the aircraft. Night training was therefore kept to the bare minimum.

'Immediately prior to the test I flew with my instructor to Croydon where I did about half a dozen practice night landings and off I went, on my own, to Lymnpe.

'The night of 15th May 1933 was a still cloudless night with a slight mist. I had endlessly studied the route and knew when and where to expect to see the lights of towns and beacons. At that time there were two lighthouse-type beacons between Croydon and the coast, basically for the use of the airliners flying between Croydon and the Continent.

'From the moment I left the lights of London behind me I didn't see a thing that I recognised. I saw towns, but never where I expected them, and I didn't see either of the aircraft beacons.

'Forty-five minutes later, still without any positive identification of my position, I knew I should be somewhere near Lympne. I should add that the flight had to be done without the help of radio direction finding, in fact without radio at all. When, according to my navigation, Lympne ought to be within sight, I circled slowly round but saw nothing that looked like a flare-path.

'Butterflies began to flutter around my stomach as I realised that I was completely lost and that somehow I must get down safely in the dark within the next two hours when I would run out of petrol. I decided to carry on for a bit in case unexpected headwinds had developed during the flight. After continuing for fifteen minutes I came to a well lit town. This I was quite unable to identify.

'I should add here that map reading by torch in the cockpit of an open aeroplane is not easy. One has only two hands and two eyes to share out the job of holding the map open without letting it blow overboard, holding the torch, holding the control column, searching the map and trying to cross-check with what one could see outside, and watching the altitude of the aircraft. If one flies 'one wing down' without knowing it, and this is easy at night, the stars can be confused with lights on the ground and vice-versa.

'Frank Payne, my instructor, was an expert at flying a Redwing with his knees and could do just about everything except bring the stick right back for the actual landing. Unfortunately I have long legs and my knees met well above the joystick so that at night I had to hold the map with my left hand, the torch between thumb and first finger and the top of the control column, or joystick as it was known in those days, with my remaining fingers. When I wanted to refold the map the aircraft just had to fly itself.

'To get back to this mysterious town, I had to face up to the fact that I was lost and somehow I had to avoid disaster with under two hours to come up with a solution to my problem. I decided the best thing to do was to turn back to London. The lights of London would provide an enormous target and once I could see London there would be no problem in finding Croydon. After fifteen minutes of flying on the return journey I was back where Lympne should be. I circled round to have one more look for this illusive airport and, to give myself the best chance of seeing what I was looking for, I switched off all the dashboard lights — both of them!

'I forgot that circling round without instruments on a dark night with no horizon is an easy way of getting out of control. The birds found that out years ago. As always happens under these circumstances, I finished up in a screaming descending spiral. I hastily switched on the lights again and continued my journey to London after sorting myself out. Unfortunately, as a result of the few minutes panic when I'd temporarily lost control of the aircraft, I absentmindedly continued on my original course.

'Surprise, surprise, when after fifteen minutes I came to a well lit town that I immediately realised I had seen before. By now, returning to Croydon was out of the question and I had to find a way of getting down, in the dark, away from an airfield, preferably in one piece. I'd given the problem of what to do if lost at night a lot of thought. I decided the answer was to land on a well lit street if such a street could be found. The Redwing had a landing speed of about 35 miles per hour and would pull up in 60 yards with no wind. Therefore if one could reach the ground, without hitting any unseen obstruction, the chance of a successful landing on a well lit street was good.

'Frank Payne, my instructor, always emphasised the need of making the best of a bad job if one got into difficulties. Panic decisions are usually bad ones. If one got lost by day, the thing to do was not land in the first big field to appear but to use ones remaining fuel to pick the best one available and to do a few low passes to check the proposed landing run before landing. All we four apprentices frequently lost ourselves yet never damaged an aircraft when landing in a field to ask the way.

'Here I was over a strange town in the dark, at about one in the morning, with about an hour's fuel left. I carefully examined every street in the town but, strange to say, I never improved on the one I spotted right away. Whilst circling the town I did notice that the sea appeared to be on the wrong side. All the streets had trams with overhead cables. Therefore the street I selected was a side street, adding the further complication that I had to be down by the intersection so that I would pass under the tram wires.

'I knew that I would also have trouble in keeping the aircraft to the centre of the road once down. The only way one could steer a Redwing on the ground was by a metal tail-skid coupled to the rudder. On a grass field this gave adequate steering but was of doubtful value on a metalled road. I would also get little or no braking effect from the tail-skid and the Redwing had no wheel-brakes.

'Having picked the street, I had to face up to gliding down it with houses passing uncomfortably close to my wing tips. A gliding speed of 55 miles per hour appears to be very slow when approaching an airfield but very fast when one is passing between houses. It must also be remembered that I had only done six night landings in my life. I have no recollections of being frightened, doubtless because I had a definite plan of action and was fully concentrating on making the best of a bad job.

'The first approach I made was too high and there was no chance of being down before the tram wires so I opened up and went round again. The next time I was at the right height but going too fast. I knew that a successful landing depended on making the approach at the correct height and speed. At about the sixth approach I felt that I'd got it right and carried on down with the houses whistling past my wing tips and everything on my body screwed up tight. I've never before or since had such an impression of speed in an aircraft.

'My landing was smooth and gentle and I touched down just where I wanted but, as anticipated, my steerable tail-skid was useless. The street was of heavily cambered cobbles so I quickly drifted into the side of the road, hitting my left wing tip against a lamp post. This swung me right round, hitting my right wheel against the kerb, snapping the stub axle off just inside the wheel.

'One of the instructions thumped into us over and over again was to avoid fire-risk by switching off before crashing. I should have switched the engine off the moment I'd committed myself to a landing so that if I hit anything the engine had started to cool and no flames were coming out of the stub exhausts. It took me a minute or two to collect my wits and realise that I was down safely and unscratched. It was then I realised that my engine was still ticking over. With a tremendous feeling of guilt and a quick glance over my shoulder to check I was unobserved, I closed down the engine.

'Until my landing the streets had been deserted but before I stepped out of the machine I was surrounded by a crowd of men and women jabbering in French. I had landed in the Boulevard des Alliés, Calais.

'As I stepped out of the machine I was thumped on the back and shaken by the hand with a chorus of 'bravos'. There were so many conflicting emotions passing through my head that I could hardly think up a word of French.

'I said 'Croydon, Croydon.'
'Ah, yes, yes.' They knew where that was.
'To Lympne.'
'Ah, yes, yes.' They knew that too.
'But I am here.' There were roars of laughter as the story was passed to those out of hearing.

'Then a couple of gendarmes appeared. I had to shake hands with them and they were told the story. I began to tire of being a one-man circus show so I collected my overnight case and the valuable Ministry barograph and took hold of one of the gendarmes by the arm saying, 'Come on, let's go.' He got the message so off we went.

'Wandering through the deserted streets of Calais, we kept meeting other gendarmes. They had to be told the story with lots of hand-shakings. Never in my life have I shaken hands with so many gendarmes.

'Eventually we reached what I assume was the Central Police Station. Here I was met by the Chief of Police. He was not in uniform so I presumed that he had been called out from his house. He was a very charming middle-aged man and he spoke good English. He took all the details he required and then he let me ring Croydon to tell them that I was alive and kicking and to ask them to inform the Redwing Aircraft Company likewise.

'I was flying their demonstration aircraft registered G-ABOK — definitely 'Okay' by me.

'Eventually, satisfied that I was an indifferent navigator rather than an undesirable alien, he said he would arrange hotel accommodation for me. Off I set again with a posse of gendarmes to a modest looking hotel. The night porter, disturbed from his sleep behind the reception desk, rallied round and stood us all a glass of beer while laughing his head off at my night's adventure. The whole party then escorted me to my room and, only when they were quite sure that it met with my approval, did they leave me with lots of wishes for a good night's sleep. Believe it or not, I slept like a log.

'In the morning a gendarme came round with an Englishman whom he introduced as the British Vice-Consul. The gendarme explained that he had no further interest in me and was handing me over to the responsibility of the British Consulate. We walked to the Consulate, a little bit of Britain where I was naturally offered a cup of tea. It was explained to me that arrangements had already been made with a local garage to have the aircraft dismantled ready for shipping to England.

'There was one snag. No funds were available for helping stranded aviators. Almost as he was saying this, a clerk came in with a draft for £25 from the Redwing Aircraft Company. As I had only about £2 in my pocket I was very grateful to Redwings for their quick appreciation of the position in which I found myself. Before I left the Consulate, to have a look at the aircraft, a bill for the equivalent of £5 came in for damage to one lamp post.

'I found that they had lifted the aircraft off the road and placed it under the shade of a statue to a famous Calais lifeboat coxswain. Many years later when I was serving with the R.A.F. out in Burma, the camp cinema was showing the relief of Calais. Suddenly, above the heads of the troops, I saw this self-same statue. Without thinking, I shouted, 'I've landed there'. The general reaction was, 'Poor fellow, he's been out east too long.'

'The garage had repaired the axle, so the Redwing was looking fine again except for the damaged wing-tip where I'd hit the lamp post. We removed the wings, a simple job with the old-type bi-planes, wheeled the fuselage on to a small cargo ship, where we stowed it together with the wings. The ship was due to leave for Dover that night.

'As already mentioned, this particular aircraft was the demonstration aircraft. In those days the Automobile Association had a very useful Flying Section. The A.A. produced excellent flying maps, helped people to prepare foreign tours and assisted at flying meetings. This demonstration aircraft had the A.A. badge on its side. While we were dismantling the aircraft, a Calais representative came up and, noticing the badge, offered his help. The job was just about complete so there was nothing he could do so, thanking him for his offer of help, I thought, 'that was that', little realising that before the adventure was over I would be very grateful to the A.A.

'My trip across the Channel by boat was completely uneventful and we arrived at Dover in the early hours of the morning. The first person to greet me was an A.A. representative offering his help. It did not take us long to discover that there was a problem in bringing the Redwing into the United Kingdom. There was no record of it ever having left the country and here I was, importing an aircraft complete with engine and instruments, all of which were highly dutiable.

'The customs officer said he knew what had happened because he had read about it in the papers but his chief would expect all the paperwork to be in order. He disappeared with the A.A. man for about two hours. They eventually reappeared with the good news that all was well. I think the A.A. man persuaded them to accept it as a 'sample'.

'I still find it difficult to believe that the total cost to the insurers, having paid my expenses, the cost of shipping it back, and the cost of repairs to wing and undercarriage came to exactly £100. Even accepting the fact that Redwings probably put in a nominal charge for repairing their own aeroplane, it still amazes me.

'Looking back on this adventure, I now realise how well I was taught to fly by Frank Payne. He was always telling us to think ahead, anticipate things that might go wrong and have a plan of action if the worst should happen. This is the only way to stop following one mistake by another through panic reactions. Mistakes of even a few minutes ago are past history. The rest of my life starts now and lets make it last as long as possible. That's my philosophy.

'Throughout my flying career I have been unhappy in an aeroplane if I know of a potential crisis for which there is no remedy. I suppose this is why I've never liked single engined aircraft. If one is flying over water or jungle or taking off towards a row of houses one can only crash if the only engine stops. I don't like single engined aircraft and they don't like me. In the small amount of single engined flying I've done I've averaged a forced landing every 150 hours, fortunately getting down safely in each case. Yet throughout the thousands of multi-engined hours I've done, only on a handful of occasions have I had to shut down an engine.'

After the Calais incident, Eric Starling had to resit his night flying test which he did this time without incident, but which didn't make half as good a story as the first effort.

One result of the publicity following the Calais street landing was the impression created on a man who was to play an important part in a later stage of Eric's flying career. It was shortly after Calais that Eric Gandar Dower diverted to Gatwick in his Puss Moth and was escorted to the club house by Eric Starling.

Gandar Dower had his own recollections of their conversation at that first meeting.

'My first pilot was met under rather unusual circumstances. Miss Brunning and I had been in Europe flying in my Puss Moth and while abroad we read of an extraordinary happening. A gentleman endeavouring to qualify for a licence got muddled up crossing the Channel and, his petrol running short, had landed in a street of Calais by lamp light and got away with it. Really one could say 'How silly!', but how magnificent.

'I flew back and I decided to go to Gatwick where there was a small private aerodrome. There a rather tall young boy greeted me.

'What's your name?' I said.
'Starling.' he answered.
I said, 'Not the ass who landed in Calais last week?'
'Yes.'
'God,' I said 'You're as mad as I am. I'm going to build an airport and start air lines at Aberdeen. Are you willing to join me?'
'Yes,' he said. And so he became my first pilot.

JE NE COMPREND PAS.

THE NIGHT WAS DARK, YOU COULD NOT SEE,
ALL FOR A TEST, IN FACT A 'B',
HUNDREDS OF LIGHTS, AND ALL OF THEM RED,
SO VERY MANY THEY WENT TO MY HEAD.
BARCLAY'S BEER?, OR LIGHTHOUSE TRIM?
NOW WHICH ONE IS THE LIGHT OF LYMNE.
I TURNED THREE TIMES AND TOOK A CHANCE
BUT THE LIGHT I PICKED WAS A LIGHT IN FRANCE,
THIS IS NOT LYMNE! IT IS TOO HARD
I'LL TAKE A CHANCE ON THE BOULEVARD,
MY LUCK WAS IN, SHE SETTLED WELL,
OH! WHAT A TALE I'LL HAVE TO TELL,
BUT WHEN NEXT I FLY BY NIGHT,
I'LL LOOK FURTHER TO MY RIGHT.

L. F. Payne. Chief Pilot
Redwing Aircraft Co.
3/6/33.

Frank Payne was Chief Pilot at the Redwing Aircraft Company. He was also Eric's Instructor.

# Se croyant en Angleterre, un jeune Anglais touriste aérien passe le détroit et atterrit... boulevard des Alliés

## Il se pose sous le réseau des câbles aériens à l'angle de la rue du Havre

Pour un homme qui a de la chance, M. Eric Allen Starlinck, âgé de 22 ans, domicilié 60, Bradhurst Garden Hamurstead à Londres W. W., 6, jeune touriste anglais, grand amateur d'aviation, a de la chance !

Il était parti hier soir assez tard dans la soirée dans son petit appareil biplan, une place, de l'aérodrome de Croydon, près de Londres, pour gagner celui de Lympne, près de la côte.

Que lui arriva-t-il ? Fut-il pris par le brouillard, ou se trompa-t-il de direction, toujours est-il qu'il traversa le détroit sans s'en apercevoir et vint, vers les minuit, minuit un quart, tournoyer au-dessus de Calais.

A ce moment, il n'était plus question pour lui de retrouver son chemin.

Pendant plus d'une heure il chercha, vainement, comme bien on pense, au-dessus des maisons, un terrain d'atterrissage et, n'en trouvant pas, se rapprocha du littoral.

Perdant de l'altitude et de la vitesse, il se trouva bientôt volant très bas, de l'ouest vers l'est dans l'axe du Boulevard International.

C'est alors, à la hauteur de la rue de Thonis à peu près, qu'il eut une chance inouïe.

Son appareil venant au sol en plein axe de la chaussée, glissa sans encombre sous le réseau des câbles aériens de sustention et de force des tramways, sans en heurter un seul, à l'angle de la rue de la Mer, pour venir s'arrêter contre le monument de Gavet devant le débouché de la rue du Havre.

En passant, il avait légèrement heurté du plan inférieur gauche le pylône dont il a décroché la plaque indicatrice et le bec de gaz voisin dont la glace fut brisée et le fût légèrement faussé.

Une roue de son train d'atterrissage heurtant le macadam avec violence, s'était brisée... et c'est tout ! Il était exactement une heure et demie.

Pour de la chance, c'était, on le voit, de la chance.

Sitôt à terre, le jeune aviateur miraculeusement échappé à la mort, se rendit, conduit par des passants, au commissariat de police où il déclina son identité et fit sa déclaration.

Puis, tandis que des agents garaient son appareil sur le terre-plein du monument, hors de la circulation, il allait prendre un repos nécessaire pour le remettre de ses émotions à l'Hôtel Excelsior, rue Royale.

Ce matin à 7 heures, l'appareil, gardé par des agents et des douaniers, était encore dans la même position.

C'est un petit biplan de tourisme, rouge et noir, à 5 cylindres, portant les marques G — A B O K.

Il est là gisant, sa patte...pardon, sa roue droite d'atterrissage cassée, mais prêt à repartir après cette réparation.

Ajoutons que le jeune touriste, avant d'aller se coucher, s'était rendu à la poste de Calais-Nord pour téléphoner à Croydon, et à Londres et annoncer sa mésaventure.

Disons en terminant que c'est M. Debief, commissaire spécial, sous-chef de service, qui s'est rendu sur les lieux pour procéder à l'enquête d'usage sur cet incident peu commun.

On se rappelle qu'il y a trois ou quatre ans, un avion du même genre, égaré dans les mêmes circonstances, tomba au rond-point de la rue de Thermes. Mais celui-là avait été entièrement détruit par le feu. Et les aviateurs qui le montaient, blessés dans la chute, ne durent leur salut qu'à l'intervention de courageux sauveteurs.

Le développement du tourisme aérien nous en promet certainement d'autres !

Médaille Militaire

Le Petit Calaisien
Mardi 16 Mai 1933

# The Fledgling Starling

Eric Allen Starling was born at 70 Boundary Road, St John's Wood, London NW8 on Thursday 28th December 1911. This was the family home, a four storey Georgian terraced house smothered with Virginia creepers.

Boundary Road was so named because it followed the boundary between two of London's boroughs, Marylebone and Hampstead. The presence of this otherwise imaginary dividing line would be brought home whenever there was a power failure in one of the boroughs, resulting in one side of the road being illuminated while the other was plunged into involuntary darkness.

Eric was the youngest child in a family of three. His brother James was his senior by four and half years while his sister Margaret was six and a half years older than him.

His father was Robert James Starling who lived from 1868 until 1942. Eric's mother was Florence and she lived from 1873 until 1931. Florence was actually a cousin of Robert's and they shared the same grandfather, James Starling (1802-1854). Robert Starling made his living from a jewellery business that also doubled as a pawn shop which was located in Great Portland Street, London W1. He had taken over the business from his father Robert Barker Starling (1835-1909).

Any observer of the pioneering of air transport will observe that many of the great names of early flying in Scotland in fact came from south of the border; people such as Edmund Fresson, Eric Gandar Dower, George Nicholson, and Henry Vallance. (The most notable exception being Greenock born David Barclay.) However a little delving into the history of the Starling family reveals that, when Eric Starling was to eventually take up a notable aviation career in Scotland, it was really a long overdue home-coming.

The Starlings had only settled in London towards the end of the nineteenth century. Prior to that they had lived at Saffron Walden in the English county of Essex. The family's presence here has been traced back to 1632 when William, George and Archibald Stirling arrived as refugees from political and religious persecution from their traditional family home at Glorat, near Lennoxtown, in Stirlingshire.

Presbyterianism was at the height of its fervour in Scotland during the early part of the seventeenth century. Episcopalians were ruthlessly persecuted and this harassment seems to have prompted the move of William, Episcopalian Minister of Baldernock, to England, followed in 1640 by his brother George and nephew Archibald.

Another migration of Stirlings to Saffron Walden followed about 1690 when Robert Stirling, the heir to Glorat, arrived at Saffron Walden. However it is conjectured that all was not necessarily 'happy families' in Walden at that time when it came to taking a consensus on religion and politics. The country had two monarchs during this period, William III appointed by Parliament and James VII of Scotland (James II of England) in abdication. Most members of the family were therefore Royalist and Episcopalian while the newcomer, Robert, seems to have been a supporter of William III and a member of the Independent Church, the closest thing that Saffron Walden had to a Presbyterian Church at that time.

As the times of political and religious turmoil passed, the Starlings by all accounts flourished in Saffron Walden. The name 'Stirling' was quickly corrupted to 'Starling' in their adopted home, although in 1903 some members of the family, in a joint deed-poll, restored their surname to the original Scottish form of 'Stirling'.

In 1771 a William Starling was born, the great-grandson of Robert Starling who had arrived about 1690, and he became land-agent to Lord Braybrooke. After marrying Ann Spicer in 1793 he gave up the land-agency and set up in business as a wine-merchant. He died at the age of 37 in 1808 and both the business and his widow fell into the care of his brother John who also became wed to Ann. John also died comparatively young in 1820 at the age of 39 and the business was carried on by William's son, James. James became a successful businessman and also served as Mayor of Walden in 1854. Later in the nineteenth century the wine business suffered when new legislation permitted licenced grocers to sell wine by the bottle and no doubt was a contributory factor in the departure of some the Starlings from Saffron Walden during this period. By 1894 there were no Starlings remaining in Saffron Walden but parish records show fifty-five as having been born in the town during their two and a half centuries of residence. In recent times the name 'Starling' could still be discerned in foot high letters above the old wine cellars in Saffron Walden's Church Street and the present owners have named the house 'The Starlings'.

And so it was that Eric Starling came to be born in London in 1911. He was originally named as Eric Allen Starling, the 'Allen' being chosen as a gesture towards an uncle who was about to die at the time of Eric's birth. As it happened, this uncle continued to live for many, many years and Eric remembers him as an enormous bluff and hearty type with waxed points to his moustache. As a child he was called 'Allen' but in adult life he reverted to 'Eric' as he had spent his childhood feeling that the 'e' in its spelling was effeminate.

Eric's father was obviously doing quite nicely in his jewellers business as Eric recollects his upbringing in that great four storey house in a world of servants and nannies which no longer exists today. Eric would have breakfast and lunch with his mother and father but to be invited to join the adult world for dinner required a very special occasion indeed.

There was a large garden behind 70 Boundary Road on to which a basement, below street level at the front of the house, opened. Eric was only two years old when the First World War started yet some recollections of that time remain vividly with him. During Zeppelin raids the family would retire to the basement of the house where French windows opened on to the garden. These were shielded with great wooden shutters and young Eric's fear throughout the noise of the raids was that these shutters would be blasted in on top of them.

Eric's father was too old to qualify for enlistment during the Great War and so he served as a Special Constable. The house wasn't equipped with a telephone and when a raid was imminent, often during the middle of the night, the family would be aroused by heavy banging at the door which resulted in father getting dressed to go out on duty while the rest of the family retired to the basement.

Eric attended kindergarten in another Georgian terrace in Belsize Road, where one day the first of a stick of bombs stripped the facade right off some nearby houses leaving all the rooms exposed to view like a giant dolls house. There was a pub at the end of the road where he recalls someone coming out to witness this devastation just in time to be caught by the last bomb of the stick.

During the Great War, one job undertaken by Eric's mother involved her taking wounded soldiers round the stables at Buckingham Palace. The soldiers from the military hospitals wore blue outfits with red ties. Accompanying his mother Eric would visit the riding school, and see the stage coaches and horses. He was particlularly fascinated to see the horses hooves being boot blacked. During this period Eric would often watch soldiers being drilled in the streets. His elder sister and a friend used to disorganise the brass bands by sucking lemons in front of them!

Just after the war an influenza epidemic ran rife in the city and his school was closed for several months as a result. Eric started his studies at the kindergarten at what would now be considered the late age of six and a half years. The headmistress was Miss Chattaway who was quickly nicknamed Miss Chatterbox. He went to a prep school called The Hall Hampstead near Swiss Cottage and close to where his father had been brought up. He recalls his father saying that he used to walk his sister home from parties because he didn't like her walking alone along the country roads at night — an area which is now in the heart of urban London. Eric himself recalls Wembley being a country stop on the Metropolitan Railway where development was just starting. Soon cranes started to appear beyond trees at the far side of the fields adjoining the railway. Signs confirmed the proclamation of the cranes that the arrival of the Wembley Exhibition was imminent.

Eric's father had been a governor of Bishop's Stortford College and Eric went to the prep school there at the age of twelve. After two years he took his Common Entrance Examination and entered the College itself. Eric's cousin, Leader Stirling, was head of school house when he was in the prep school and, as the elder cousin, he considered it his duty to take Eric out for a walk on Sundays which Eric recalls as 'being all rather painful'. It was only in conversation many years later that Leader Stirling revealed that he held a similar sentiment towards these Sunday afternoon strolls. Leader Stirling took his unusual forename from his mother's maiden name — she had been a 'Miss Leader'!

It was during a winter term at Bishop's Stortford that Eric caught chickenpox which resulted in his enthusiasm for flying. Confined to the college sanitorium, he found himself next to another student by the name of Geoffrey Sanders who was fanatical about flying. They became great friends and spent many hours making model biplanes out of wooden match boxes. Sanders came from a very wealthy family. He had already learnt to fly before leaving the college and he soon had his own aeroplane.

Eric left the college at seventeen and a half without much academic distinction — his only good subjects had been science and mathematics, and perhaps just as important in the college environment, he left without any great sporting distinction either. He claimed to be too flat-footed and round-shouldered making his power to weight ratio too lacking in harmony for him to be a good sportsman. Shining in neither realm, his departure from college was accompanied by a bit of an inferiority complex.

In later life however Eric fathomed why all his friends seemed to be such great academic achievers while he did so poorly in exams and he has come to the conclusion that intelligence was not his short-coming but that he was served by a very poor memory. There was undoubtedly much truth in the first part of this once upon a time but I find the second part of the reasoning a little difficult to accept from a man who now has such a store of wonderful recollections.

Having parted company with the college Eric embarked on a three year apprenticeship with an advertising agency, Saward Bakers in Chancery Lane, London WC2. The apprenticeship also included night classes but Eric did not see himself as an advertising man. Colleagues would discuss things like the balance of a composition but he freely admits that he could rarely see what they would be getting at in their discussions on such abstract subjects.

This period did however allow Eric to pursue his enthusiasm for flying and he joined the Herts and Essex Gliding Club. He describes the activity here as 'gliding down the side of Hog Hill, Hainault, slightly under control'.

'The gliders were very primitive affairs which we all used to break — and repair — periodically. Flight was achieved by catapult take-offs using two enormous lengths of rubber material with four people at either end with a ninth member of the team holding the tail of the flimsy craft.'

When he was half way through his three year advertising apprenticeship Eric was one day summoned into his boss's office. 'I'm chucking you out.' said his boss. 'You are only interested in flying. If you are given the choice of an advertising magazine and an aviation magazine you will pick up the aviation one every time. Dismissing you is the kindest thing that I can do for you.'

It may have seemed a harsh ultimatum but Eric's employer tempered it with an undertaking to help him in any way he could and to arrange some introductions. One of these was a luncheon appointment with Lord Brabazon. As it happened Eric had to make his next step, the one that took him into the world of flight, on his own initiative.

## An Abridged Starling Family Tree

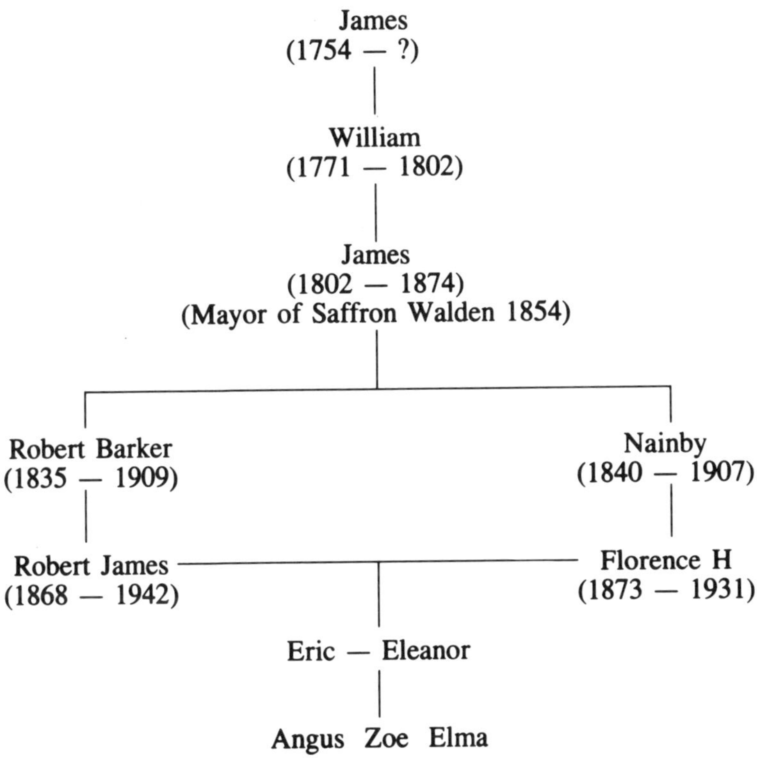

# 'Take Off' with Redwing

Having left the advertising world behind, Eric found himself with time on his hands while he re-assessed his future. Aware of the situation, his father despatched him on a course of typewriting lessons which he wisely speculated would be a skill that would always be some kind of an asset. To observe the array of computer keyboards in Eric Starling's home today, it is immediately apparent that it was a skill that has served him well over the years.

At this time Eric applied for a short service commission and while awaiting a response to this he one day saw a newspaper advertisement placed by the Redwing Aircraft Co. Ltd. The company had been purchased by an American, from the General Motor Car Company, called Besner.

In acquiring the manufacturing facility for the Robinson Redwing aircraft at Croydon, Besner had dreams of turning this side-by-side two-seater single engine bi-plane into the Ford motor car of the air. Besner sought to recruit four English ex-public schoolboys to train as apprentices, the objective being to give them a thorough grounding of the product and then 'set them on to an unsuspecting world to sell his aeroplanes'.

Eric wrote to Redwing and obtained an interview, which he attended in the company of his father. This was conducted by another American called Roland Darling who was running Redwing on Besner's behalf. All parties concerned measured up to scratch and Eric's father enrolled him at a fee of about £20 per month.

During the two years that followed, Eric served in all aspects of the Redwing operation, learning both how to manufacture the aircraft and how to fly it. The apprentices were given the prototype Redwing I, registration number G-AAUO, to maintain under supervision, and if they damaged it they had the job of repairing it under supervision too. It was in this aircraft that Eric took his first lesson on 11th August 1931, and also went for his first solo flight seven hours and forty minutes of flying time later, on 8th October 1931.

Just before Eric joined Redwings, a number of trainees had been sent over to England to qualify as aircraft engineers for the Indian Air Force. They passed their exams with flying colours, but when they returned to India and started maintaining aircraft, they were found to be doing an extremely poor job. It transpired that they were all good at learning things by heart and could tell the examiners exactly what should be done when things went wrong, but they were very lacking in practical experience. The Indian Government complained bitterly and as a result no subsequent engineer student was ever allowed to sit the necessary exams until he could prove that he had had two years practical experience at aircraft maintenance.

As Eric's apprenticeship included both flying and engineering training, he spent his time in the factory when not in the air. Close by Redwing's factory was a shed belonging to the Aircraft Disposal Co. After World War I, they had bought up hundreds of surplus engines and aircraft from the government. Eric recalls that there were hundreds of wingless and engineless Avro 504K fusilages stacked on end. There were also dozens of Rolls Royce Eagle Eight engines. This engine was the forerunner to the Merlin. Redwings bought from them a 504K fuselage for £5 and an Eagle Eight for not much more. Eric and his fellow students spent hours dismantling and reassembling the engine and they were taught the difficult job of rigging the 504K fuselage. There were so many cross bracing wires, each tightened with a turnbuckle, that it was a miracle if one finished up without a twisted fuselage. He remembers one foreman who had started in the shipbuilding industry. He said that there, one's personal tool kit was treated with great respect. It usually contained specialist tools made by their owner for a particular job. Eric still has tools that he made when with Redwings. This foreman said that if someone was away sick or for any other reason, his personal tools were carefully collected and nothing was missing when he returned. It was a lesson that Eric later noted seemed to be absent from the training of the Air Force engineers.

Airfield control wasn't so organised at that time. There was a purple faced pilot by the name of Muir who flew for one of the small charter companies at Croydon. Air France were flying a machine which looked rather like a large Dragon Moth but with an open cockpit on the top. Muir was landing at the same time as Air France. He saw no reason why he should give way to a French airliner so they approached and landed more or less in formation. As they taxied in the Frenchman could be seen standing up in his open cockpit gesticulating at Muir. Fortunately no one could hear what he was saying. On another occasion when Air France were trying to land in the early morning mist, the pilot overshot through the fence, across the Sutton road and finished up with his nose against a house. The cockpit was level with the first floor bedrooms so someone must have awakened with a French airman looking through his or her bedroom window.

Few people know that only a few years before Eric joined Redwings, KLM had a woman pilot flying into Croydon. Surely the first female airline pilot. She was Lady Heath, the first woman in the world to get a commercial pilot's 'B' licence after World War I. In 1936 she said in an article that she had flown 76 types of aircraft and gliders and had logged 2,670 hours. Additionally she had flown 70,000 miles as a second pilot with KLM. She died in 1938 after falling down the stairs of a London double-deck bus. What an undignified ending for an outstanding woman pilot!

An interesting aircraft flying into Croydon during Eric's time there, was the giant Junkers G38 with four Jumo diesel engines. The engines were inside the leading edge of the wings where there were also two cabins for three passengers between the inboard engines and the fuselage. It had an all-up weight of 52,000 lbs and cruised at 129 m.p.h. It carried 34 passengers between Berlin and Croydon. Eric remembers going across to the tarmac during his lunch hour to watch it land. He was most amused to see that while the passengers were disembarking at the back, a circular trapdoor opened up on the underside of one of the wings and three crew members dropped out on to the grass and walked off to the coffee bar.

Imperial Airways' 'Jumbo' was the Handley Page 42. It had an amazingly short take-off run and Eric has seen it take-off from the tarmac in a high wind. Captain O.P. Jones was one of the captains at that time. Many of the Imperial Airways pilots were a bit rough and ready but Jones was one of the first of their pilots to put some dignity into the job and his uniform and appearance was always immaculate. On one occasion he was flying with a new second pilot straight out of the R.A.F. As they started their take-off run, the second pilot shouted, 'Whoa! Whoa! The starboard engine is packing up!' As Jones pulled back the throttles, he said gravely, 'Whoa! Whoa! indeed. This is an airliner, not a horse and cart'. In 1931 — to Eric — O.P. Jones was a God. He remembers seeing him, a fine athletic figure of a man, swimming with his son in the Redhill baths. When Eric met Jones many years later at a party, he reminded him of this. With a sad look on his face, Jones said, 'That son of mine is now in a wheel chair after catching polio.' Eric is forever thankful for the luck which had pursued him all his life. Unlucky pilots kill themselves the first time they do something stupid. Lucky pilots like Eric, scare the living daylight out of themselves but survive to be wiser pilots.

There were coveys of partridges on the airfield which seemed quite unafraid of the aircraft. When the wind was blowing from the east and aircraft had to taxi to the far side of the field for take-off well away from the control tower, Frank Payne, Eric's instructor, would take a twelve bore gun with him. If there was a suitable covey on the far side he would taxi close to it. The drill was, that the moment he fired his gun he would switch off the engine so that the tower would think that it had backfired and stopped. Frank would jump out, pick up the dead partridge, swing the prop, get back in and fly off.

After a couple of months at Croydon, two things happened. The first was that it was decided that airliners and flying instruction didn't mix and all flying instruction at Croydon stopped. Secondly, Redwings decided to move their factory to a farm owned by John Howie at Colchester where operating costs would be lower. John Howie's family originated from Ayrshire. Two fields were knocked together to form an airfield and the factory and local flying club house were erected in a corner. It was here that Eric took his first solo flight at the age of nineteen. This meant that Eric Starling qualified to fly an aircraft before he was licenced to drive a motor car.

In reality there was no driving test at that time. Drivers just paid five shillings and got a licence. Eric's first ventures behind the wheel of a car were totally informal. Redwings had an old car that everyone drove. The stores were kept at a different part of the aerodrome from the factory so whenever he was asked to collect something heavy Eric took the car without explaining that he had never received driving lessons. One day he was asked to collect someone at the station and off he went for his first drive on the open road although both his employers and his passengers were kept in blissful ignorance of this fact.

John Howie was also an accomplished pilot. On one occasion he had to make a visit to his lawyers in Scotland as he had been having trouble with contagious abortion with his cattle which was resulting in some financial problems. So he flew up to Scotland and invited Eric to join him for the flight. They followed the east coast to Newcastle then crossed to Carlisle. Just before crossing the border they landed in a field at Longtown and refuelled the aircraft at a nearby garage. Aircraft at that time ran on standard car petrol. On trips like this, an essential piece of equipment was a funnel protected by a chamois leather filter to eliminate any possible water or dirt.

They then continued to Kilmarnock where they landed on the farm of a sister of John Howie. Mr Howie then suggested to Eric that he take a maid at the farm up for a flight, since she had never been in an aircraft before, while he went into town to conduct his business.

Eric didn't like to let on that he had only done five hours solo flying and shouldn't really be carrying passengers. So he took her up and circled Kilmarnock. However, the aerial viewing of the town completed, Eric realised that he hadn't taken much interest in the landmarks for the farm when they had landed since he had been the passenger. It now occurred to him that he wasn't sure which farm he had taken off from. He circled round until he spotted a farm with a large field. 'Doesn't your farm look lovely from the air?' he said to his young passenger. Fortunately she agreed and so they landed without the maid ever being aware of the predicament.

From Kilmarnock they flew to the farm of John Howie's mother in Moscow (a village in Ayrshire) followed by Eric's first visit to Renfrew Airport. Eric recalls flying across the River Clyde where the Queen Mary was a rusty hulk in John Brown's shipyard in its early stages of construction, and indeed not yet with a name.

Back at Colchester, his flying instructor was called Frank Payne. The students became known as Payne's Fireworks and when Frank Payne got married the students were to do a fly-past at the church. Frank didn't have too much confidence in this manoeuvre being completed without an aircraft coming through the church roof, but they all came safely to rest in a nearby field and the young pilots then joined the wedding celebrations.

Eric was to meet Frank Payne again many years later during the 50th anniversary celebrations at Gatwick Airport. Having survived forty years of flying, Eric was delighted to have this opportunity of thanking Frank for teaching him to fly so well.

During the Redwing two year apprenticeship Eric gained his Engineering 'A' and 'C' Licence and also his Pilots 'B' Licence, the equivalent to the present Commercial Pilots Licence.

It was at this time that Eric began to lose his inferiority complex because for the first time he felt he was being a success at what he was doing. For example he found that whenever a V.I.P. visited the factory, he was always the one to be chosen to perform a little flying display 'to show off Redwing's pilot apprentices'.

The examinations which had to be undertaken in order to gain the Pilots 'B' Licence included a night flying test. Another test involved the student flying to Northolt where the Royal Air Force undertook training. Redwing had bought Gatwick aerodrome by that time and Eric flew across a fog-bound London at 3,000 feet to take his exam, the weather at both Gatwick and Northolt being clear.

The programme here included half an hour of general flying tests followed by a forced landing. Eric flew to Northolt in a Redwing but he then had to take his test in a De Havilland DH60 Gipsy Moth. He had never flown in this aircraft type before and the unfamiliarity coupled with stage-fright resulted in a fairly poor performance. For the forced landing the examiner pulled back the throttle leaving Eric to make a half-hearted pass over an unsuitable field. The result was that the examiner suggested that Eric put in another twenty hours flying and then try the test again.

Depressed, Eric was heading back to home base in his Redwing when, over Twickenham, his engine started to peter out right above Twickenham football ground. A few days earlier another apprentice had encountered problems with the same aircraft and had to make an emergency landing on Epsom Race Course. On that occasion the instructor had come to his aid and found nothing wrong with the engine. The poor apprentice was accused of trying to get out of doing that particular assignment since it was known that he had had a heavy date planned for that night.

Eric, aware of this incident, nursed the spluttering aircraft away until, at Hook, half way between Esher and Ewell, he eventually spotted a likely field for an emergency landing. However as he got lower he saw that the field was divided up into squares with wire netting. Eric aimed for the first square, came in a bit too fast, but landed nicely in the second square. Fortunately one of the attributes of the Redwing was its ability to take-off and land within only sixty yards.

It was a source of some amusement to Eric, having fluffed his mock forced landing during his test earlier that day, to see that his instructor had some difficulty in putting his aircraft down in this field when he came to assist. Again the stricken aircraft started without difficulty and the instructor flew it back to Gatwick while Eric followed in the instructor's machine. Next day the aircraft was taken up for tests and at 3,000 feet the engine again started to cut out. A new fuel system had been fitted to this machine and it was now discovered that it was vapour-locking.

Flying as a passenger on a flight from Glasgow to London Heathrow during April 1988, the descending aircraft passed over Twickenham. It amused Eric to see that one of the stands carried on its roof the advertising slogan 'Next time fly KLM', advice which he would have considered highly appropriate had it been there in 1933.

The only time that Eric has watched the Oxford & Cambridge Boat Race was from the air on 1st April 1933 and then he had to land in order to find out who had won. 'Thirty eight aircraft were circling the course including two giant Imperial Airways Handley Page 42s. None of us were in radio contact with anyone, but the Air Ministry had given out an instruction that left-hand circuits must be done over the course. Such was our casual attitude to controlled flying in those days.'

Under normal flying conditions field landings were regular occurences. There was a road house called 'Old Barn' at Hildenborough which had a field big enough for landing light aircraft. Eric recalls, 'If you flew there they gave you a free lunch. According to my logbook I flew there with a friend to scrounge a free lunch on both the 5th and 30th of August 1933 in the Redwing G-ABOK.' Eric also availed himself of a similar facility on 23rd July 1933 at a road house called 'The Ace of Spades' at Kingston. These road houses considered a pilot's meal to be a good investment since they drew additional patrons attracted by the presence of the parked aircraft.

On 4th November 1933 Eric visited his old college at Bishop's Stortford and landed on the college playing field. The groundsman was not amused when he crossed the cricket pitch scoring it with his tail skid. On a subsequent visit Eric decided not to incur the groundsman's wrath further and chose to land across the less hallowed ground of four rugby pitches.

The Redwing apprentices were taught two procedures to follow if they became lost. The first was to fly low and slowly past a railway station to read its name and the second was to land in a suitable field and ask the way. For this they were taught to make several low passes before landing and check that there were no hazards on the selected landing run. If there was 'ridge and furrow' they landed diagonally across the furrows. It was also just as well to make sure that there was someone about who might be able to give directions. In one instance a pilot landed on a large field beside a road where a man was walking. His appearance was that of a farm yokel. 'Where am I?' asked the pilot. 'I don't know' — 'Well, where does this road go to?' — 'I don't know' — 'Where does it come from?' — 'I don't know' — 'Damn it man, don't you know anything?' — 'No. But I ain't lost!'

N.M. Russell, Redwing's Chief Pilot, had a similar experience. He could speak French fluently and was flying to Paris to demonstrate the Redwing. Finding himself lost, he landed in a field with a farmhand working in the next one. No amount of waving would get this man to come over to him. So, much against his better judgement because he was wearing his best shoes and clothes, he picked his way across a very wet and muddy ploughed field. He then asked his wearabouts in impeccable French only to find that he was addressing a Polish immigrant who didn't understand a word of it.

As previously mentioned, the Redwing Aircraft Company had purchased Gatwick, at that time a grass field aerodrome. Immediately north of Gatwick was Gatwick Racecourse. When meetings were on, many jockeys used to arrive by air. They often used to give the Redwing staff tips of which they took full advantage. Many of the tips were good and they did well from them. This was the only occasion when Eric took to backing horses but he did come out on top revealing that he made 'not a lot, but enough'.

Another nearby neighbour of the airfield at Gatwick was the main London to Brighton railway line, a feature that has been fully utilised by the modern airport. The engine noise experienced in an open cockpit aircraft is sufficient to obliterate any other sounds. However Eric encountered an exception on one occasion when flying a Redwing into Gatwick. He was throttling right back and gliding over the railway line when a train passing below blew its whistle. Eric nearly jumped out of the aircraft when this unexpected shriek emitted from below.

Redwing ran a flying school at Gatwick and the apprentices had to take turns to act as 'duty pilot'. This involved looking after visiting aircraft, refuelling them and taking their pilots and passengers to the club house to be fed. On one occasion the Prince of Wales arrived in his Puss Moth flown by his personal pilot, 'Mouse' Fielden. Fielden was later Commander of the Royal Flight for many years and on this occasion Eric had the pleasure of swinging his propellor, self starters then still being a novelty.

It was during a shift as 'duty pilot' that Eric Starling was to meet someone who was to be of much greater significance when Eric Gandar Dower, in the company of Caroline Brunning, arrived from Paris in his De Havilland DH80A Puss Moth. Eric's night flying test, having been full of incident, naturally came up in their conversation giving Gandar Dower much amusement.

Late in the summer of 1933 Eric had completed his apprenticeship period with Redwing, and had also successfully concluded the tests for his Pilots 'B' Licence. This enabled him to join the Reserve Air Force Officers which paid him a retainer of £25 a year. But at this point flying jobs were not plentiful and he continued at Redwing for a short time doing testing and demonstration flights. Soon, however, Redwing called Eric in and told him that they felt that they could keep him no longer. By good fortune a letter arrived during the same week from an old school friend to say that he was starting a flying circus, and would Eric like to join him?

As for Besner's dream, one of the aircraft manufactured by Redwing's was sold as far away as New Zealand. Another aircraft set off for South Africa piloted by airwoman Mrs Keith Miller. But it crash-landed at Cotonou in Dahomey. This was not enough to establish an enthusiastic market across the world and only about a dozen Redwings were actually built. It did not, after all, become the Ford motor car of the air.

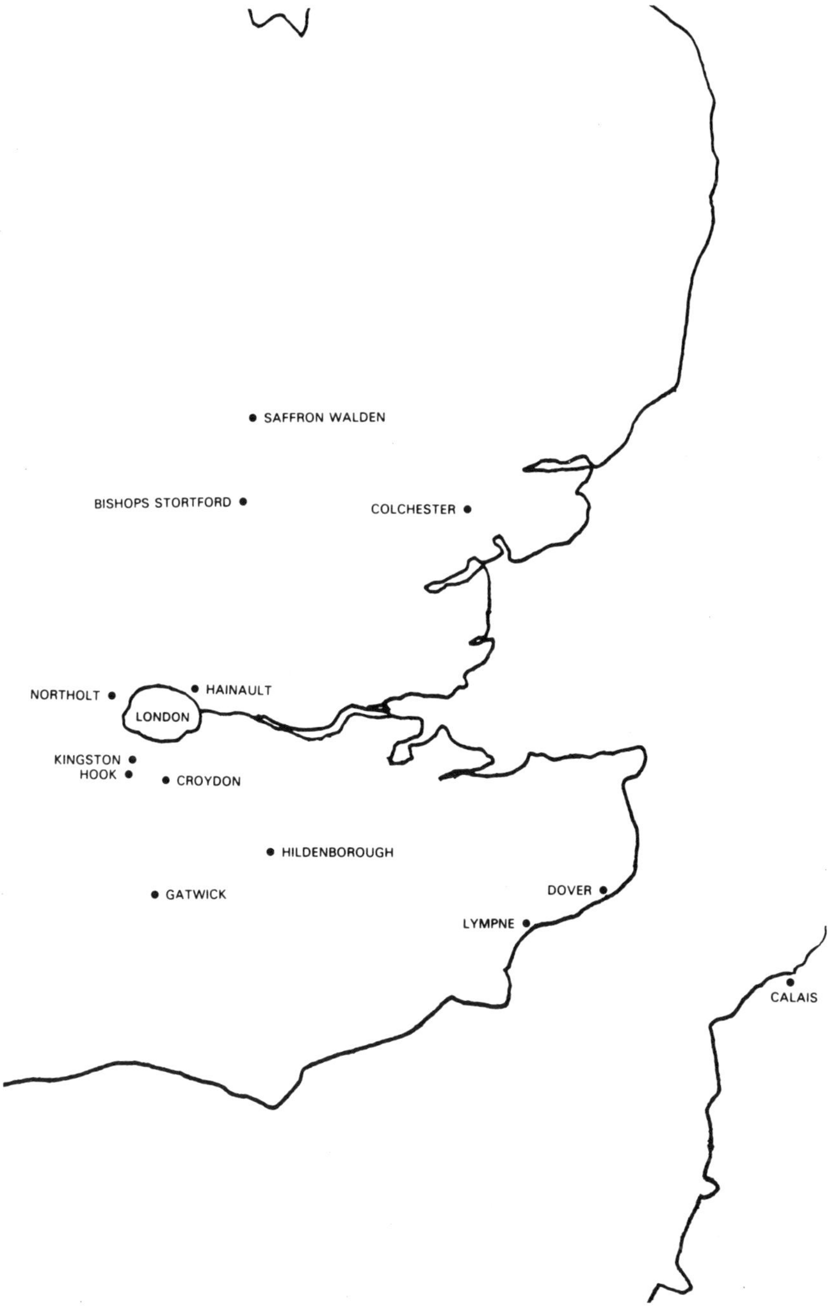
SAFFRON WALDEN
BISHOPS STORTFORD
COLCHESTER
NORTHOLT
HAINAULT
LONDON
KINGSTON
HOOK
CROYDON
HILDENBOROUGH
GATWICK
DOVER
LYMPNE
CALAIS

# On the Edge of Flying Folly

Eric spent the winter of 1933/34 undertaking his first task for 'The Flying Fair', the name that his former school friend, Dixie Gerrands, had given to his enterprise, although it was properly titled as Aviation Developments Ltd! Here is how it developed.

At the wheel of an old car that the flying circus put at his disposal, Eric toured the southern half of England seeking out potential sites for the venture when it went on tour.

In locating potential sites the tactic was to ask in a local pub if any flying circuses or barnstormers had visited the area previously and where they had operated. Using the experience of previous aerial displays made the task of locating landing grounds much easier and many fields used by Allan Cobham, undoubtedly the most famous barnstormer, were adopted by The Flying Fair.

Each landing site to be used had to be licenced by the Air Ministry. A commercial pilot holding a 'B' licence, was responsible for doing this and it entailed marking an Ordnance Survey map with the various obstructions, landing runs, distances, etc and then sending a signed copy to the Ministry. A temporary licence for joy-riding was then issued by the Air Ministry.

The flying circus eventually opened up at Southend on 29th March 1934. The team consisted of eight or nine aircraft with Eric mostly at the controls of the familiar Redwing, on these occasions G-ABDO. Providing 'ground support' was a Fun Fair which would provide diversions for the spectators in between displays.

The participants in the flying fair made up their stunts as they went along. These were dictated by the skill of the pilot and the versatility of the aircraft. The Redwing wasn't very suited to rolling but Eric would put it through loops, spins and stall turns developing a taste for 'crazy flying' which he confides has never since left him.

Unlike the Tiger Moth, the Redwing did not have enough power to climb out of a loop. Instead Eric would do the stunt in reverse. He would start the loop a few hundred feet up and dive out of it across the field. Not realising that he was deliberately delaying recovery from the loop, the spectators would be on their toes wondering whether Eric would pull out of the loop in time.

A stunt created by the team was called 'Cutting the Toilet Roll'. The end of the toilet roll was loosened and it was thrown from the aircraft. The loose end would act as a parachute and unravel the rest of the roll as it floated through the air. The stunt was then to fly through the toilet roll as many times as possible before it reached the ground. Hill, the announcer, would then inform the crowd that if anyone wished to recover a piece of the paper (not explaining that it was toilet paper) the pilot would be only too pleased to autograph it! The daily autographing of toilet paper was not a finale to the stunt that Eric appreciated.

Another gimmick was called 'Bombing the Car'. One of the team, Daly, was a former clown and he would circle the field driving an old open-topped Austin car. He would use his former skill to full effect by acting the fool, while Eric became equally skillful in hitting this target with paper bags of flour from above. Walls Ice Cream were permitted access to the field for ten per cent of their sales. One day when the car wasn't available, the ice cream vendor was persuaded to cycle around the field on his ice cream tricycle. The poor soul was soon white from head to foot and spent the rest of the day cleaning up his tricycle instead of selling ice cream.

The technique developed by Eric was 'skip bombing'. He would approach low and slowly then drop the flour bag over the side just before reaching the target. The bag would bounce and hit the car on the side covering it with flour. Using this method Eric became deadly accurate and scored more hits than misses. As he came over one day, the clown in the back of the car held an umbrella high above his head and Eric knocked it out of his hand with the aircraft wheels. He didn't believe that he could possibly have done this until presented with the damaged umbrella for inspection.

Bursting balloons was another diversion. The hydrogen filled balloons were a target for the aircraft propeller with a hit being marked with a loud bang. Eric recently witnessed both this and paper cutting at a flying show near New York where he admitted to a feeling of frustration and conceit as he was convinced that he could have done it much better. And he thought their crazy flying was 'wet'.

The Flying Fair included a parachutist called Ben Turner. Upon arrival at a town, Ben's first job was to find a hall where he could pack his 'shute after use. It was usually Eric's job to take him up for the jump. The cockpit was too small for Ben and his parachute so he used to sit on the edge with his feet hanging out over the side. All the extra drag didn't help the performance of the Redwing and it used to take quite a while to get up to the necessary height of 600 feet. When Eric arrived at the desired position for the jump, Ben used to shout 'hi-di-hi' and over he went. Another pilot at the Flying Fair was Ted Jordan who was joy-riding in a Spartan three-seater. He used to describe himself as an elderly gentleman who flew in the last War. He later became Flight Manager-Jersey for B.E.A. Jordan used to bait Turner by saying there were only two reasons for being a parachutist — you hadn't the brains to realise what you were doing or you were too lazy to do real work. A local paper recorded the visit of the Flying Fair to Swindon as follows:

*Snappy Display*

*The Flying Fair, which attracted scores of local people yesterday afternoon and evening, is perhaps one of the snappiest displays seen in the district, and the programme is carried through on non-stop lines.*

*The fair will be in Swindon until Sunday. At present there are only three 'planes at the field but it is expected that two more will be arriving today.*

*Captain Jordan, in his red three-seater Spartan machine, gives a wonderful display of aerobatics. Passengers are taken up during these displays.*

*A hair raising exhibition of crazy flying is given by Mr. Starling in the two-seater Redwing, which has side-by-side seating accommodation, during which 'turn' he makes his machine fly in a drunken manner.*

*Parachute Jumps*

*Two competitions are held daily, for which free flights are awarded — a speed and height test, and these are carried out in the small Klemm monoplane. It has a nine cylinder 40 h.p. engine and has an extremely low landing speed.*

*Owing to the high wind yesterday it was not thought advisable to carry out the two proposed parachute jumps, although the parachutist expressed his willingness to do so. If the wind is not too high today there will be one jump this afternoon and another this evening.*

From Southend the Flying Fair zig-zagged its way across southern England giving aerial displays at various towns and villages. Moving from one field to another in their aircraft, it never took the team long to move on to the site of their next performance. However they were usually well behind the Fun Fair whose seasoned travelling folk were well-practised at packing up early in the morning, moving on and setting-up again with customary precision.

Eric struck up a good friendship with many of the Fun Fair people. A motherly figure came in the form of Ma Peake who would greet the flyers, when they belatedly arrived at a new field, before a table at her caravan laden with slabs of bread heavy with jam. Eric recalls taking a fancy to her daughter, Jane Peake, whom he flatteringly recalls had a voice like a chain saw but a terrific personality!

Jane ran a stall at which Fair goers were induced to try their skill at throwing pennies into squares to win prizes. The flyers used to keep performing until the light finally failed. They would then make a bee-line for the pub before closing time. However they were frequently waylayed by Jane Peake who would furnish them with pennies and then award prizes if the dashing aviators got anywhere near a square. They would leave laden with prizes which they returned at the back of the stall before dashing to the pub. Jane obviously considered this ploy good for business.

A gypsy family with the Fun Fair could neither read nor write and Eric was often called upon to read their letters for them. They told Eric how, on one occasion when travelling, they had come to a sign post and entered into a discussion as to which way they should turn, no one admitting what they already knew — that not one of them could read the directions. The mother was the most 'literate' member of the family who could read — peoples fortunes! After much thought, the old father one day plucked up courage to ask

Eric's Redwing G-ABOK awaits repair of its axle beneath the silhouette of a lifeboat coxswain in the Boulevard des Alliés, Calais, while a watchful gendarme stands guard.

Eric aged eight years — 1919.

The facade of 70 Boundary Road, smothered by Virginia creepers, that was Eric Starling's childhood home.

Florence and Robert Starling, Eric's parents.

A first taste of flight for Eric came at the Herts and Essex Gliding Club where Eric would fly on a Dagling Glider down Hog Hill, Hainault, 'slightly under control'.

Eric (in the cockpit) with fellow Redwing apprentice, John Newman, at Castle Douglas. It was in this Redwing, G-AAUO, that Eric flew his first solo on 8th October 1931.

At the controls of Redwing G-ABOK above Gatwick Racecourse August 1933.

Passengers board Imperial Airways' Handley Page HP42 'Hengista' at Croydon Aerodrome during the early 1930s.

Frank Payne, who taught Eric to fly, in leather overcoat and goggles, with two other Redwing apprentices, Thomlinson and Newman.

Payne's 'fireworks' — Newman, Pritchard, Thomlinson and Eric.

John Howie, the Flying Farmer, on whose land Blue Barns Aerodrome was laid out at Colchester.

Eric if he would 'take him up'.

A high-light of the Flying Fair for many daring spectators was to go for a five shilling flight which would maybe last for five or ten minutes. But spectators could also elect to go as a passenger for the full aerobatic sequence on payment of £1. Eric would strap his passenger in beside him and he would then delight in saying to him 'Isn't this a stupid way to earn a living?' while flying upside down, the passenger no doubt going through combined feelings of exhilaration and sheer terror. Eric's Redwing was the cheapest aircraft to operate so if there was no-one game to pay to participate in the aerobatic display, Eric would be the pilot to put on the solo performance.

At Watford the field was on a slope requiring the aircraft to land uphill but take off downhill. This meant that landing and taking-off aircraft were facing each other but both landing and taking off aircraft had a clear view of each other and it worked well.

At Kenilworth Eric had his second forced landing. Here one took off over a valley with a row of trees on the other side. Eric would take-off down the valley, go up the other side and then around behind the trees. This would enable him to circle round out of sight and come roaring back to the field from the opposite direction where the crowd were still concentrating their vision towards the trees in search of the aircraft.

On one occasion however Eric had just reached the bottom of the dip when he saw smoke coming from his engine and oil splattering his windscreen. A glance at his oil pressure gauge indicated that he had no oil pressure at all. Hoping that the engine would keep going, he scrambled up the other side of the valley, went behind the trees, but didn't appear again from any direction. The other aircraft took off in search of Eric to find that he had landed safely in a ploughed field where good fortune had the furrows running in the right direction for a comfortable emergency landing. The incident had been caused by the oil pipe to the oil pressure gauge breaking close to the engine and oil was squirting out from there. In reality Eric would have been able to fly round the circuit and land normally but he wasn't of course to know that at the time.

During one stunt of upside down flying Eric experienced a shrill 'ping'. One of the cross bracing wires on the centre section of the Redwing had snapped. He had to turn the aircraft around the right way again which put a lot of strain on it. Fortunately there was ample built-in redundancy of bracing wires and he rolled the right way up without further mishap though with a few anxious moments.

At Great Malvern the field lay in the lea of the Malvern Hills. There was a row of trees at the end of this field too. On one occasion the hills created a down draught as Eric was taking off. His heart missed a beat as the Redwing just made it across the tree tops ahead of him.

Many of the fields used by the Flying Fair were very small. On one occasion the flyers had come in over the boundary hedge as low and slowly as they dare and this would make the crowd cheer enthusiastically. Once or twice Eric came in too low and he could hear the hedge scraping the undercarriage. He kept quiet about this until he heard Ted Jordan, veteran of the First World War say, 'This field is bloody small. Several times I've put my wheels through the hedge.' Eric then admitted that he had done the same. Ted's reaction was, 'That's fine provided the hedge doesn't have a strong fencing wire running through it.'

The Flying Fair had a German-designed Klemm Swallow, a low-wing monoplane with a very low landing speed. It had a nine cylinder 40 h.p. Samson engine with a metal propellor. The engine ran so smoothly that, at anything less than full throttle, it felt as if the engine had stopped. The instruments were in French and Eric was never sure what the correct speeds were. The aircraft performed well in high winds so Eric was using this aircraft type on one particularly gusty day and, when coming in to land, he realised that the field was in fact receding. The wind was blowing him back faster than he was flying forward so he had to open up a bit to catch up with the galloping airfield.

The economics of the Flying Fair were put under strain at an early stage. But Eric and his colleagues were having such a good time that they decided not to draw a salary but settled instead for a share of the takings at the end of each day. 'The pilots were expected to go as near to the edge of flying folly as they could. And they did, emerging unruffled and unscathed. It wasn't as if we were being paid danger money,' recalls Eric, 'for the pay was a trifle paltry, beer money you might say.'

Eric bought a small tent which he erected at each field. The life was totally carefree because food cost nothing as an ex-ship's cook had been installed in a single deck bus converted to a kitchen. The thrills in the air, and a little revelry in the local pub after sunset, were all that the team were seeking for their labours. However when they reached Exmouth during June 1934 the enterprise finally went broke and the team made a hasty departure from the field there, before the farmer impounded their aircraft in lieu of payment. Eric flew his Redwing from Exmouth to Gatwick on 19th June where presumably it had been hired from its manufacturers.

When Eric had left the Redwing Aircraft Company he had written to Eric Gandar Dower seeking a job as a pilot with the airline that he was planning to form at Aberdeen, and which he had mentioned during their brief encounter at Gatwick after the Calais incident. To Eric's disappointment, a reply to his letter had not materialised, and so he had joined the Flying Fair.

However, by a strange coincidence, a letter had arrived from Gandar Dower only one week before the Flying Fair had run out of money. The letter said, 'Meet me at the Royal Aero Club with a view to a job.'

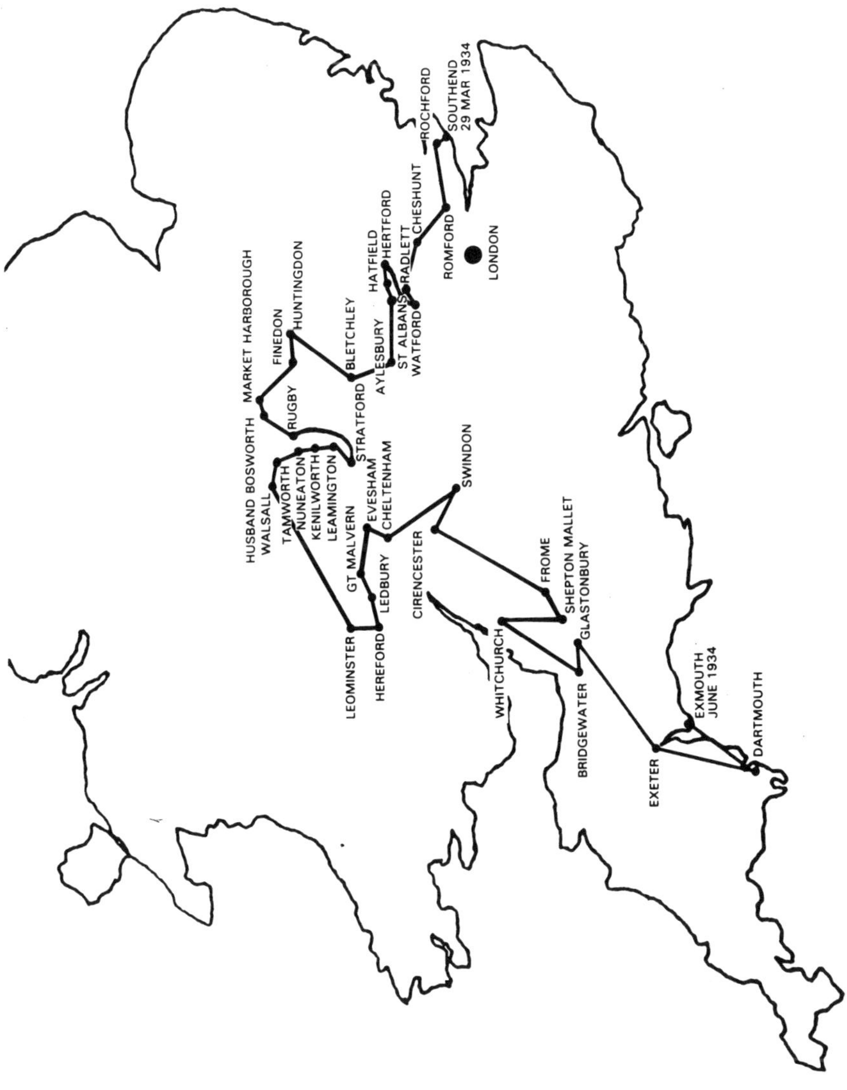


The Fying Fair 1934

# Air Lines at Aberdeen

Eric Leslie Gandar Dower was born in London on 2nd July 1894. His father was a partner in a tea broking firm. The family were of very substantial means but Eric, one of five brothers, spent his early childhood as a stranger to them all being brought up by his grandmother. This no doubt gave rise to an air of independence that was a characteristic that stayed with him throughout his life.

Eric Gandar Dower did not reach adult life with ambitions of blazing the skies as a pioneer of new air services. He did have a passion for flying but in his adolescence this was shared with an equal fervour for the theatre. Early inspiration was supplied by Louis Bleriot when he crossed the Channel in 1909 and by Tryggve Gran's 5 hour 40 minute crossing of the North Sea from Cruden Bay to Stavanger in 1914. Gandar Dower made a point of travelling to see both events.

Early attempts to become a flyer had been thwarted and with the arrival of the First World War Gandar Dower found himself in the Army. He didn't take to it and so spent a one month period of sick leave taking flying lessons. This move became his ticket for a transfer to the Navy where he was soon made a Flight Sub-Lieutenant.

After the War his flying was put into cold storage as he pursued an acting career. He spent the next few years almost constantly 'on tour' and by 1920 at the age of 26 years he had founded his own theatrical company and was showing his own productions. However he was aware of a deadline that was getting ever closer. His father had died in 1919 but his disapproval of Eric's acting career lived on in the form of a clause in his Will that prevented his being a beneficiary of the Family Trust if he was still on the Stage by his twenty-eighth birthday

Eric Gandar Dower therefore decided to divert his energies and to redeem himself in the family's eyes by going to Cambridge to take a degree in Law. During this period he participated in many activities including the Cambridge University Amateur Dramatic Club. During his final year he met Robert Cowell Smith from Banchory in Aberdeenshire and it was he who sowed the seeds in Eric's imagination of starting a flying school in the north-east of Scotland.

Having graduated with an M.A. in Law in 1926 Eric Gandar Dower entered into a period of socialising, particularly with the aviation fraternity at the Royal Aero Club. It was during this time that he met his life-long companion, Caroline Brunning, although that was in different circumstances and Caroline came from a background that was in total contrast to the affluence of the Gandar Dower family.

In 1929 Gandar Dower purchased his first aircraft, a Blackburn Bluebird IV, and this was followed a year later by a de Havilland DH80A Fox Moth. In 1931 he entered the King's Cup Air Race and participated in this premier event annually until 1935. In 1931 he decided also to visit Aberdeenshire and pursue the idea of a flying school that his friend Robert Smith had proposed. In the company of another great friend, aviator Angus Irwin, the Bluebird landed on a 15 acre field adjacent to the village of Dyce on Sunday 22nd March 1931.

It was to be some time before this first landing was to be followed by the formation of a regular air field at Dyce, but in October 1933 the task of converting stony farm land into an airport had begun. By January 1934 a hangar had been erected and four different companies had been registered. Aberdeen Flying School Ltd would train aviators and run the airfield while Aberdeen Flying Club Ltd would attend to the recreational needs of flyers. Aberdeen Aerodrome Fuel Supplies Ltd provided the ground support for aircraft, while Aberdeen Airways Ltd was the most ambitious project which aimed to develop air transport for the carriage of passengers, mail and freight. The hasty establishment of these plans was prompted by the presence of Edmund Fresson who had established Highland Airways at Inverness the previous year and was already looking to Aberdeen as an extension to his operation.

By the summer of 1934 Gandar Dower was getting ready to open his airport at Dyce and so summoned Eric Starling to meet him at the Royal Aero Club in London for an interview. Eric got the job but he learned from Caroline Brunning some time later that it had really been a *fait accompli*. Gandar Dower remembered Eric from their meeting at Gatwick after his Calais street landing and, on receipt of Eric's letter shortly after, he had remarked to Caroline, 'Now there's the sort of fellow we want to use; with plenty of enterprise and initiative. When we're ready to take on pilots, see that we get him!'

When Gandar Dower signed Eric up he required him to get various aircraft types on his 'ticket'. These included the de Havilland DH84 Dragon, the Short Scion and the de Havilland Fox Moth. He also required Eric to undertake an instructor's course. Gander Dower said that he would make arrangements with de Havillands and Short Brothers for the Dragon and Short Scion to be put on Eric's ticket but he would have to make his own arrangements for the Fox Moth.

Eric knew the club instructor at Southend and they had a Fox Moth, so Eric made arrangements with him to take the test to get this aircraft added to his ticket. The routine for this consisted of three 'light' landings and three landings with a full load. Eric was given a briefing on the single control aircraft and then off he went. 'When I came round to land, the wretched thing wouldn't sit on the ground. It just floated right across the field. Round I went several times but I still couldn't get it to land. After about the third attempt I realised that the throttle wasn't coming back properly so I gave it a good bang this time to bring it back and landed very nicely and taxied round. As I was taxying round to do my second circuit the instructor rushed out and waved me in.

''What's the trouble?' asked Eric. 'Get out of that bloody aeroplane.' came the reply. 'I'll sign anything but get out of my aeroplane!' So Eric was signed up for the Fox Moth but in fact never flew one again as Gander Dower never acquired this type of aircraft. One wonders why?

Gandar Dower had Klemm Swallow aircraft in his flying school and on 18th July 1934 Eric went to Hanworth to get the Swallow added to his ticket. The next day Eric went to Rochester were he did his test for the Short Scion and had it added to his licence. This aircraft carried five passengers and was to form part of the Aberdeen Airways fleet. This was followed by some local flying in Short Scion G-ACUV which Gander Dower had purchased.

Eric did his flying instructor's course at Broxbourne under the watchful eye of 'Grieve', later to become Wing Commander Grieve, Chief Flying Instructor, at the Coastal Command School that was to crop up later in Eric's career.

On 27th July Eric set off in the company of Angus Irwin for Aberdeen in the new Short Scion. The duo flew via Brough and Cramlington with Eric at the controls during the intermediate leg of the journey. The Short Scion was a stop-gap aircraft brought in to replace Gandar Dower's first de Havilland Dragon which had been the casualty of an early accident. As it happened, the Scion was to serve the fledgling airline faithfully for five years.

On the following day Dyce Airport was officially opened. The field was still quite small at that stage but Gandar Dower had already started up the flying school. The flying school aircraft shared the field with dozens of nesting plovers.

The plovers were no doubt some of many spectators to witness the range of activities laid on that day to mark the official opening of Dyce. These included a variety of competitions with Edmund Fresson, Gandar Dower's protagonist at Highland Airways, carrying off a silver cup and £30.

Eric's contribution to the proceedings was to carry joy-riders in the brand new Short Scion G-ACUV. Consistent with all of Gandar Dower's early aircraft it carried an incongruous purple and yellow colour scheme. These were the colours of the Cambridge University Dramatic Society and were perhaps Gandar Dower's way of expressing a last defiant gesture by perpetuating his final connection with the theatrical world. Eric undertook a total of seven trips that day, all fully laden, thereby giving the first taste of the new airline to thirty-five intrepid passengers.

At the time of the opening, the field at Dyce was very stony. But during extensions the following year Gandar Dower had the field 'Hunterised'. Hunter was a specialist in levelling and smoothing grass airfields. Eric first met him when he was 'Hunterising' Gatwick airfield. One of his tricks was to drag large heavy girders behind a tractor. This not only scraped the ground level but also compacted it.

The plovers nesting on the field were casualities of this process but one nest close to Gandar

Dower's house on the perimeter caught his affection and instructions were given that it should not be disturbed. This solitary plover remained on its nest mounted on a column of earth while all the others had been levelled off.

Two days after Dyce had been opened Eric was given a plane-load of passengers resulting from a frivolous wager proposed and lost by Gandar Dower over a business luncheon. His forfeit was to treat his guests to a flight over Loch Ness for the purpose of 'monster spotting'. Eric set out across the Grampian Mountains with his passengers which included Cathcart-Jones who had accompanied Ken Waller in the original Comet on its flight to Australia. They had a pleasant day for flying over Loch Ness but of course without seeing the elusive monster.

Returning home, they were just clearing the Grampian Mountains when one of the Scion's two engines stopped over Muir of Fowlis. This aircraft would not carry on with one engine so Eric had to force-land. He located a nice field which had a slight downhill slope but this didn't matter because it also had a standing crop which would help as a brake. The force-landing was completed successfully and later the engineer arrived from Dyce. He ran some checks and the engine seemed to be running satisfactorily again.

Gandar Dower dissuaded the game passengers from continuing by air and Eric took off in the empty aircraft from the adjacent steep downhill grass field to complete the short journey back to Dyce. He climbed to 600 feet when again the engine stopped and he had to repeat the force-landing in the same field. Now he had no passengers to act as ballast so the crop tripped the aircraft up on to its nose. As the tail came up Eric was ready to walk on to the roof of the aircraft but fortunately the tail fell back to the ground before it came to a stop.

The damage caused to the nose of the aircraft had been minimal but repairs had to be done the following day. The cause of the engine failure was discovered to have been water in the petrol. There was a fuel gauge on the top of the wing and the glass on it had been broken and rain had been getting in. He arrived back at Dyce to find Gandar Dower talking to Captain Lamplugh. Lamplugh was the Lloyds insurance assessor, a very powerful man in the flying world since pilots had their own personal insurance rating. Therefore if a pilot had been accident prone his insurance rating became so high that he would be unable to get a job.

Eric thought he had done rather well during the previous day, successfully completing two forced landings and there being minimal damage as a result. However Gandar Dower, putting on a show for Lamplugh, tore Eric off a monumental strip with a tirade of 'Do you think I would have let you fly that aeroplane if I'd known that the fuel gauge was broken?' and so he went on. Eric was dumbfounded thinking that Lamplugh would slap a massive surcharge on him and he would be finished before he had started. However as soon as Gandar Dower had left the room, Lamplugh drew Eric aside and said, 'I've been in this flying game for a long time and I think you did very well. Don't you worry!'

Eric breathed a sigh of relief and gained a great respect for 'Lamps' as he was widely known.

In August Eric continued to expand his experience with a visit to Hatfield on the 8th to do the three light and three heavy landings on de Havilland Dragon G-ACRO and so get this type added to his licence. The Dragon was an eight-passenger bi-plane and it was to form the backbone of the Aberdeen Airways fleet. Three days later at Hanworth Eric took a similar test for the de Havilland Puss Moth on Gandar Dower's personal aircraft G-ABLS. He then piloted Gandar Dower via a meandering route to Dyce that included Stapleford (Herts), Rochester, Hedon (Hull), Cramlington (Newcastle) and Macmerry (Edinburgh). Dyce was reached on 15th August.

On 16th August a newly acquired de Havilland Dragon G-ACAN arrived at Dyce from Hillmans Airways and Eric spent the next three days practising different manoeuvres under varying conditions in the airliner. At a later date G-ACAN was named 'The Starling' by Gandar Dower in honour of his young Chief Pilot.

Flying assignments were of a spasmodic nature at this time. One of these included a charter with the same Puss Moth on 27th August carrying a Dr Glass and a Mr Stewart via various points to Hull. On the return journey Eric's dislike of single-engined aircraft reaffirmed itself when, flying through bad weather only 200 feet above St. Andrews Golf Course, the engine packed in. He had just flown over a nice big field so did a quick about turn and effected a smooth landing. The incident had been caused by a rocker arm breaking.

On 4th September 1934 Eric flew the Short Scion on a trial flight for the proposed scheduled service from Aberdeen to Renfrew. At Renfrew they picked up John Sword, Director of Midland and Scottish Air Ferries. Sword was also Western Manager for the Scottish Motor Traction Co. Ltd. and had been under some pressure to run either buses or aircraft, but not both. Sword had therefore decided to sell his airline and he joined the party which continued on to the Midland & Scottish destinations of Campbeltown and Islay. Gandar Dower was obviously evaluating the operation but in the event he decided against taking it on.

On 10th September 1934 Eric flew a party of inaugural guests, including Gandar Dower and Viscount Arbuthnott, in the Scion from Aberdeen to Renfrew. The return journey was preceded by a flight over the 'Queen Mary' in its final stages of construction at Clydebank and a lunch hosted by the Provost of Glasgow, Mr A B Swan, in the then plush (but now demolished) Kenilworth Hotel in Glasgow's Queen Street.

The regular service commenced the following day. The schedule allowed for two round trips each day and Gandar Dower had invested in a bus to convey passengers to Dyce. However these proved to be a rare species. In the first week of operations three passengers had been carried contributing a total of £8 15/-while the operating costs totalled a staggering £450. While some of the services operated with uneventful regularity, other flights in both the Short Scion and de Havilland Dragon were disrupted by mechanical problems

Eric awaits the next customer to be game enough to join him in his Redwing for some crazy flying.

Two of the gang take a break at the chuck tent during a visit to Leominster with the Flying Fair. The bus served as the kitchen.

Curious bystanders are attracted to Aberdeen Airways' brand new Short Scion. This aircraft carried five passengers and pioneered the airline's first scheduled routes.

Early mainstay of the Aberdeen Airways fleet was the De Havilland Dragon Moth. Here Eric poses before 'The Silver Ghost'.

The trim lines of the Klemm Eagle, one of the trainers used by the Aberdeen Flying School Ltd.

Crowds throng Dyce airfield during a 1930s air display.

Dyce airfield and village as it was before the Second World War.

An early shot of DH-84 Dragon G-ACAN at Renfrew. Here it is still in the colours of its previous owners, Hillman Airways, and yet to be named after Eric.

and weather conditions with diversionary landings being made at points such as Forfar, Leuchars and Perth. An encounter with a pilot who had flown with Allan Cobham's Flying Circus resulted in Eric getting all the fields they had performed from in Scotland marked on his map. This gave him a wealth of information on suitable diversionary fields.

The flights operated without the aid of radio, making mist over the hills a regular hazard in poor weather. The most reliable means of navigation proved to be the railway line and Eric frequently followed the silver highway on his way to and from Renfrew. The highest point on the railway was Gleneagles. Eric came to know the Gleneagles Hotel intimately, having approached it from all angles and passed over it from great heights right down to nought feet, yet he has never had occasion to step inside the hotel through all the years.

The Puss Moth was used when there were no passengers to be carried, and it was this aircraft that Eric had to put down in a field at Forfar when he ran out of daylight. Caroline Brunning, who excelled at handling Gandar Dower's finances but knew nothing about aviation, reprimanded Eric with a terse 'You must not force land!' The east coast of Scotland being extremely changeable in its weather, this instruction was remembered by all the pilots with a degree of amusement for some considerable time.

On 24th October Eric diverted to Abbotsinch, site of the present Glasgow Airport, to make his first landing there. Renfrew, two miles away, was fog bound. It was at this time that Gandar Dower decided that it was proving futile trying to compete with the railways over the Aberdeen-Renfrew route and the service was abandoned after seven weeks of operation.

In 1934 the feeling was already around that a war was a distinct possibility and Eric foresaw that, as a professional pilot, if this happened he would very likely be dragged into it whether he liked it or not. Commercial pilots could get a commission in the Reserve Air Force Officers. A retainer was paid and once a year members would be called upon to take part in two weeks flying exercises. As Eric was regularly flying he was rarely called upon to fulfil this part of the contract. He did however attend a fortnights reserve training at Brough, near Hull, in February 1935 and this was carried out on Blackburn B2 Trainers. The B2 was an all-metal aircraft, a much improved version of the Blackburn Bluebird. By the time war did break out, he had little knowledge of the Air Force but had the advantage of already being a very senior flying officer.

Radar had yet to arrive and at that time the Territorial Army were needing regular practice with searchlights. Flying backwards and forwards across a ten mile leg at 10,000 feet for the searchlights was boring for Eric — but remunerative for Gandar Dower. Flares made from oil drums burning cotton waste were used for landing in the dark. Saturday was the usual night for this exercise so it was not uncommon for the pilots to fly their searchlight routine dressed in dinner jackets and then dash into town as late arrivals to any party that was on the go. Searchlight flying was done for Dundee as well as Aberdeen and the cause for concern here was whether Dyce would remain open so that the journey back in the dark, without radio, could be successfully accomplished.

Gandar Dower's first personal aircraft, the Blackburn Bluebird G-AASV, was kept at Dyce. Eric remembers it as a horrible aeroplane that was under powered and over ruddered. It would spin easily if carelessly handled and Eric blames the Bluebird for being responsible for more accidents involving amateur pilots that any other type at that time. On 20th February 1935 Eric force landed the Bluebird at Banchory. The engine was making strange noises and losing power. This was caused by a plug falling out. Another single-engined aircraft that didn't endear itself to Eric!

That winter there had also been a bad accident in England where a pilot, untrained in blind flying, had elected to crawl under low clouds until he ran out of visibility and pranged his aeroplane. After that a blind flying course or 'Instrument Rating' became the law for Commercial Pilots. Gandar Dower therefore sent Eric on a 'blind flying' course which he did at Cramlington during March 1935. Eric had to share a bunk with a merchant seaman who was learning to fly while his ship was laid up. The man had a constant twitch (St. Vitus's dance) which fortunately subsided during the night. Outside, rabbits dug up coal on the airfield, it being so near the surface.

On 12th April 1935 Eric first flew Gandar Dower's Klemm Eagle. It was a modern low wing monoplane with a retractable undercarriage and carried two passengers making it equally suited to joy-riding as well as pupil instruction. It was painted in Gandar Dower's 'ghastly purple and yellow colour scheme' and Eric noted that it took a good length of runway for take-off when fully loaded.

When not attending courses during the winter of 1934/35, one of the activities in which Eric was involved was the Aberdeen Flying School and this generated a reasonable flow of patronage. Among those pupils coming forward to try their skill in this new challenge was Dr Agnes Thomson. She was Aberdeen's first woman doctor and she was about fifty when she decided to take flying lessons. Dr Thomson always ran a party at the annual Aberdeen Beach Ball and always invited Eric to join them. Dr Thomson had a daughter whom she kept pushing in Eric's direction, but he was more interested in the Woman Editor of The Press & Journal who came to report on the Ball and then join the party. She was Eleanor Castel and she was later to become Mrs Eleanor Starling.

Since abandoning the Aberdeen-Renfrew scheduled service, Gandar Dower had been observing the apparent success that Edmund Fresson was having with his services to Orkney operated by his Inverness based Highland Airways. Gandar Dower decided that routes across water offered a greater chance for success than those over land since he could compete more effectively against slow and often uncomfortable shipping services than against an already highly developed and fast rail service. And so Aberdeen Airways looked to the Northern Isles for their next move.

Air Lines at Aberdeen 1934

# Battle of the Northern Isles

When Gandar Dower decided that he wanted to launch scheduled flights to the north, Fresson already had a year's start on him. On the one hand Gandar Dower could observe and learn from the experiences of Highland Airways but on the other hand his competitor had already established a presence in a relatively sparse market and had first pick at the best landing sites.

Gandar Dower had control of his airport at Dyce and refused Fresson access to it for his services. Therefore he could expect no welcoming open door from the Highland Airways chief at the airports under his control — Hillhead at Wick and Wideford Farm at Kirkwall. Gandar Dower had in fact extended a belated olive branch to Highland Airways with offers ranging from co-operation to merger. However Highland Airways had just purchased a new aerodrome at Kintore for their Aberdeen services, having been repeatedly denied access to Dyce, so they felt that they should persevere independently having made this investment. In May 1935 control of Highland Airways passed to United Airways, an English company, so scope for further co-operation talks seemed to fade and battle commenced.

On 27th April 1935 Eric flew the Short Scion to Thurso in search of a suitable field from which to operate. In the party were Gandar Dower and Caroline Brunning. The first field investigated had been used by Cobham's flying circus but they weren't too happy with it. After some circling, two adjacent fields were selected and these alternated depending on wind directions. Later a hedge was moved from one of these in order to enlarge it so as to take aircraft from varying directions and this became Thurso's Clardon Aerodrome.

The Scion then proceeded to Kirkwall. This visit did not produce any likely alternative to Fresson's field and so four days later another visit concentrated on Stromness, the other main town of Orkney. Eric spotted a nice field close to the Bridge of Waith, two miles north-east of Stromness, and came in to land on it. They had been in this field for only a short time when a man joined them, introducing himself as a member of the local Chamber of Commerce. He recommended that, as Fresson was serving Kirkwall, Stromness would be a good place on which the newcomer should base its operation. Telephone wires had to be buried to make the field safe for use and this modification remained fifty years later although it no longer served as an aerodrome after the Second World War. On the following day Eric flew on to Shetland to make his first landing on the site of the future Sumburgh Airport.

As the hasty preparation continued for the new services, Eric flew a Mr Cameron of the Air Ministry to Thurso to inspect the field they had selected there. During the flight Cameron asked Eric if he could borrow the map of the route. Finding that Eric wasn't carrying any maps, he said, 'Don't the regulations say that you should have a map to cover your route?' Eric quickly produced an envelope, drew an outline of northern Scotland marking in Aberdeen, Inverness, Wick and Thurso, and said 'There's my map!' Cameron laughed and commented, 'Well, the rules don't say what kind of map you must have.'

On 27th May 1935 the first northern service was inaugurated with Eric flying the eight passenger de Havilland Dragon G-ACAN non-stop from Dyce. There were seven empty seats on this historic flight although on the return leg the load was greatly improved with two passengers! Two round trip flights were to be operated six days a week with the first week of operations carrying a total of nine passengers.

At Stromness the main aerodrome was at Howe Farm but until the telephone wires were buried this field could only be used for north/south take-offs and landings. In the interim a neighbouring farm at Cumminess, but on the opposite bank of a wide channel linking the Sound of Hoy with the Loch of Stenness, had to be used when winds required an east/west run. The traffic 'boy' was therefore kept on his toes guessing which field the inbound aircraft would select so that he could have the passengers assembled at the right place.

Gandar Dower hadn't totally given up his plans of operating to the south and on 4th June Eric used the Dragon G-ACAN to inaugurate a Dyce-Edinburgh (Turnhouse) service. However if the Stromness service got off to a slow start this new route gave even less cause for optimism. There were no passengers on board this first round trip service and grand plans to tie in with a North Eastern Airways Edinburgh-London service were to meet with frustration. The RAF station at Turnhouse did not have radio and therefore the route presented the same problems as those encountered en route to Renfrew. The route ground to a halt after only a month. But a pseudo-charter from 'Turnhouse' presented itself as a fitting gesture of contempt for a finale, as Eric recounts.

'This was a Puss Moth charter and I was to fly General Sir Ian Hamilton from Montrose to Nairn where he was to inspect the Territorials camping there. This was on 16th July 1935. At that time Montrose was a disused ex-1914/18 War aerodrome. I shouldn't have used it for a commercial flight and I see in my log that I have entered it as having operated from Turnhouse. The General was a very entertaining man and his little joke was that although he was wearing his full regalia, he said he didn't bring his sword in case he got struck by lightning. I was going to land him at Inverness and let him motor to Nairn.

'On the outward flight we passed inland of Aberdeen and I pointed out Bennachie, the most prominent mountain in the area. 'Where's the Gardie?' he asked. I hadn't a clue. Remember I was fairly new to Scotland. He then began to sing the song which goes, "Where the Gardie runs by the banks of Bennachie." The Gardie is so insignificant that we were unable to spot it from the air.

'When we reached Nairn, I circled the camp and noticed that the field next door to the camp was big enough for a Puss Moth to land and take-off. He readily agreed to my landing there.

'When the time came to leave, I saw to my horror that the troops had been paraded in this field and I was expected to take-off down the line. It was a very bad idea to my mind, apart from the fact that I wanted to give myself the maximum take-off run, that is between opposite corners. Instead, they were persuaded to line up on the other side of the hedge.

'Shortly afterwards he sent me a photo of himself in full uniform and with a hand-written message down the side, "For E. Allen Starling from Ian Hamilton — in memory of a record and dashing descent by an Inspecting General upon a Parade in a Puss Moth — 16 July 1935". It is one of my treasured possessions.'

The Orkney service was gradually becoming established and this was aided on 11th June when the Stromness-Thurso service was opened, an early start delayed no doubt by a shortage of envelopes carrying a map of northern Scotland on them. The fifteen minute hop across the Pentland Firth caught on very quickly and soon became a popular alternative to the often rough ferry crossing from Stromness to Scrabster.

In anticipation of further growth Eric went to Hatfield on 9th July to be licenced to fly the de Havilland DH89 Dragon Rapide. The Rapide was faster than the Dragon and Eric returned with the first of several to subsequently join the Aberdeen Airways fleet, G-ADDE which was appropriately named 'The Aberdonian'. Lack of radio was always a handicap but on 12th October 1935 Gandar Dower took delivery of Dragon G-ADFI, 'The Silver Ghost', and this aircraft was fitted with radio and additional blind flying instrumentation.

The sea crossing between Thurso and Stromness was fairly short but the stretch from Aberdeen to Caithness crossed a large expanse of water and this caused the Coast Guard some concern since aircraft fell within their responsibility once over the water. If an aircraft fell into the Moray Firth it would be the Coast Guards lot to fish it out again. This was before the Ministry installed Aircraft Ground Radio and aircraft were fitted out with radio telephony tuned to the International Shipping distress frequency. This enabled Eric and his fellow pilots to call up Wick shipping radio station to advise them when they were leaving the coast at Portsoy for Lybster, the landfall in Caithness, where they would check-in again.

Wick shipping radio station supplied vessels at sea with bearings for which they levied a charge of five shillings. The captains of trawlers fishing in the same area would select one of their number to order a bearing while the rest listened in and then split the cost. The Coast Guard station at Wick agreed to supply the service to aircraft without charge. However Eric found this courtesy of dubious value since they didn't appreciate their speed of travel and took ages to give pilots a bearing. The fifty mile sea crossing took about thirty minutes and Lybster was often in sight by the time the bearing came through on the radio.

When Aviation Ground Radio stations were installed radio communications were conducted on 826 meters and this could be picked up at the lower end of the Long Wave on a standard radio set. People would often use this to establish whether Eric was flying, if he was on time, and when to expect the arrival of friends who were passengers on his aircraft. The staff in the huts at the various airports used this too, in the absence of proper radio facilities. If Eric was on a special trip that might involve an overnight stop he would tell his landlady to listen to the radio at a certain time so that he could let her know whether to expect him home that night. Aberdeen was out of earshot of the nearest official radio station at Kirkwall which was unable to pick up these informal messages.

As the radios were not always that effective, a parlance was established to counteract inaudibility. If the ground station couldn't understand a message it would interject with something like, 'We presume you are on course from Portsoy to Lybster. Switch on twice if that is correct.' On the way in from the north the last call could just be picked up by Kirkwall when the aircraft reached Fyvie and where the message would jokingly be, 'We are passing Fyvie. One, two, three, four, Fyvie.'

Most of the pilots had learned Morse Code at some time or another and carried Morse keys. They couldn't read Morse but they could send it slowly. If the radio telephony wasn't working well a message would be sent in Morse and the ground station would reply verbally. This worked well except when there might be a relief operator covering holidays who was unaware of this arrangement and would respond in rapid Morse that the pilots couldn't transcribe.

Eric became used to making decisions on whether the weather would permit flight. It was often a case of 'no see, no fly' and flight at zero feet became an art. There was never radio at Aberdeen during Eric's time there and southbound flights making landfall at Portsoy often hit poor visibility. This left the option of either flying along the coastline to Dyce or following the railway line. Eric frequently selected the railway line and found that it rarely failed him. The coast route could often be followed to the mouth of the River Don but with visibility preventing the river being followed upstream to Dyce and requiring diversion to a landing ground at Seaton. During the pre-war years radio facilities did not exist at Stromness, Thurso or Dyce (where they were resisted since their presence would also mean granting access to all airlines, including Highland Airways). Direction finding equipment was installed at Wick, Kirkwall and Sumburgh during 1936. Yet despite the lack of radio facilities the airline managed to maintain a regularity of about ninety-eight per cent.

On one occasion Eric had been unable to follow either the coast or the railway and had to land in a field not far beyond Portsoy. On one side was a hospital while to the other there was a cemetery. 'So all contingencies were covered,' recalls Eric. 'When the weather cleared I continued to Dyce where Gandar Dower enquired where I had landed. 'Bloody Mire' was my reply — since that was the name of the farm.'

When the infamous east coast haar descended it sometimes took a couple of days to lift. This would frequently create clear conditions at sea but a blanket of fog over the land. Eric encountered this on 12th September 1935 and had to land at Ackergill Tower near Wick. The coast was perfectly clear but Wick was about three fields inland. Eric was able to fly inshore for about one field then all would be lost to sight and he had to veer back to the coast. He then located the large field at Ackergill Tower where he landed and waited for the weather to clear.

When the haar was encountered further south Eric used a field at Buckie for diversions where he would wait for an improvement in the weather that would permit him to continue to Aberdeen.

On 8th November the clouds were very low and visibility poor making flight inland impossible and coast crawling difficult. Eric decided to head along the coast of the Moray Firth towards Lossiemouth where he knew that there was a long strip of sandy beach. Eric takes up the story.

'I was flying south in Dragon G-ACAN. The weather was deteriorating when I finally found the south coast of the Moray Firth and, by flying at nought feet, could just follow it along but with no hope of getting inland. I knew that there was a long beach at Lossiemouth and made for that.

'When I eventually found the sands, the tide was in. However, I was committed because I did not have enough fuel to divert back to Kirkwall and I was in fog with visibility a doubtful 200 yards.

'I landed with one wheel in the water and one on dry sand. I knew that the wheel on the dry sand would dig in but, as I had hoped, the wheel in the water kept going and, instead of going on my nose, I swung at right angles, finishing up with the tail wheel in the water. My four commercial traveller passengers leapt into the water with complete disregard for their trousers and shoes, to help me swing the tail out of the water.

'We could not move the aircraft, so there we had to leave it. We had no idea whereabouts on the sand we had landed because the sand just disappeared in the fog in both directions. We walked inland, soon finding a farm. I arranged transport for my passengers and phoned Ben Brown, the engineer at Dyce, to come out to help.

'By the time I got back to the sands, the fog had lifted. There was no Dragon Moth in sight. I had just decided that the aircraft must have been washed out to sea when I looked over my shoulder and there was the Dragon Moth, high and dry up among the dunes with a small crowd around it.

'As I approached, a coastguard met me and asked me if I was the pilot of the aircraft. He said that, when the fog had lifted, he saw the machine and, as it was abandoned below the high-water mark, it was his duty to pull it out of harm's way. He got hold of several fishermen who pulled it up and put it on wooden planks.

'The Dragon is only made of wood and fabric and the wings in particular are very easily damaged by applying pressure in the wrong places. So I expected to find a lot of damage done by the willing workers pushing and tugging.

'To my surprise, I found that not a penny-worth of damage had occurred. On complimenting the coastguard, he said that they had found the undercarriage was solid so had tied ropes to it and pulled on them. I then realised that they were all seafaring folk who were used to working with boats which were similarly constructed with weak points and areas able to withstand high pressure.'

Also during November Eric had flown south in Gandar Dower's Puss Moth G-ABLS on a special charter which he recalls as not being particularly successful. His passenger was a local M.P., Sir Edmund Findlay, who was in pursuit of a substantial proportion of his electorate who had moved from the north-east of Scotland to Great Yarmouth to work on the fish-gutting there. Eric was to fly Sir Edmund to Norwich but he underestimated the required flying time and ran out of daylight. He was reluctant to continue flying in darkness without radio and so he landed on a field near Sutton Bridge, on the Wash, where he arranged for his passenger to be taken on to Great Yarmouth by taxi. Eric arranged to meet Sir Edmund the next day at Norwich and found a hotel for the night.

When Eric returned to his aircraft the following morning he arrived to find that he had been preceded by a Customs Officer. At that time it was not uncommon for entrepreneurial flyers using light aircraft such as the Puss Moth to do a bit of smuggling. However, Eric managed to convince the Customs man that he was just 'an enthusiastic and slightly inexperienced commercial pilot'. On the return trip he ran out of weather at Edinburgh and landed at Macmerry where he had to stop overnight before being able to continue to Dyce. Sir Edmund Findlay parted company with Eric here and made his own way back to Aberdeen. Eric never heard what Sir Edmund thought of him as a pilot.

On 3rd December Eric landed at St. Margaret's Hope on the island of South Ronaldsay incorporating it as a stop on the Orkney service. The island is now connected by causeway to the Orkney mainland but in 1935 this was an important link for the islanders.

The trans-Pentland service between Thurso and Orkney continued through the winter and this busy route continued to be reasonably profitable, particularly when shipping was disrupted by rough seas. However, the Aberdeen-Orkney service stopped for the winter on 16th December.

Eric recalls being based at Thurso that winter. 'I opened the Thurso to Stromness service on 16th December 1935 with on-demand stops at St Margaret's Hope, South Ronaldsay and Longhope, Hoy, continuing until 16th March 1936. I stayed with my engineer, whose name I think was Reed, at the Pentland Hotel and each day we flew to the Orkneys and back. Our Dragon Moth sat in the open and when the weather got really cold the Gipsy Major engines were brutes to start. It was all hand propeller swinging in those days. We would have to go out to the field in the morning to make quite sure that we had the engines started and warmed up for the service. The petrol we used was obtained by filling four gallon drums from the local garage and poured into the aircraft through a chamois leather filtered funnel. It had to be non-leaded fuel. One day I tried, as an experiment, leaded fuel. The experiment was a success and we had no further trouble starting. However when the engines went down for overhaul at a later date, lead deposits were found and I got a mild telling off by the Ministry.

'There was a fellow by the name of Pat Agnew staying at the hotel. He was engaged to a Dutch girl, a Baroness MacKay. At that time the head of Clan MacKay was a Dutchman. In due course she came over with a Dutch girlfriend to visit Pat. They arrived late, after I had gone to bed. I heard a lot of laughter and giggling going on outside my bedroom. When I got up in the morning and opened the door to collect my shoes — in those days shoes were left outside for the 'boots' to clean — I found my enormous size 12 shoes down the passage beside a neat pair of women's shoes.

'The Dutch girlfriend, Nan Tonken, was a very attractive blond and being young and shy in those days I found myself completely tongue-tied in her presence. She suddenly left the room and came back with a letter which she handed to me. It was signed by Albert Plesman, K.L.M.'s supremo, and stated that Miss Nan Tonken, who had worked for K.L.M. for many years, had left them to study international aviation and would people give her every assistance. That certainly broke the ice and I couldn't go wrong after that.

'Thurso had a swimming club which made a point of 'swimming the New Year in'. They tried to get me to join in without success. I never did like plunging myself into cold water. I remember that at boarding school we had to have a cold bath each morning. Just a plunge in and out. I never did like it and to the day I left I would dodge my cold bath if I thought I could get away with it.'

On 15th January 1936 a new airfield was opened at Quanter Ness, just north-west of Kirkwall thus hotting up the competition with Highland Airways. On 29th January Eric made a special trip from Dyce to Stromness carrying a 400lb load of Daily Express newspapers. This was a special edition marking the funeral of King George V.

On 2nd February 1936 Eric flew his first air ambulance flight and this was also the first air ambulance operated by Aberdeen Airways. His patient, an elderly man, was flown in 'The Silver Ghost' (Dragon G-ADFI) from South Ronaldsay to Stromness en route to hospital in Kirkwall. On board were a Mr Williams and Alf Cormack, the 'boots' of the Pentland Hotel at Thurso, who had come to assist because of high winds. Compared with modern aircraft, these Dragons had enormous wings which caused two problems in high winds. It was very difficult to turn across the wind because, having the wheels well forward, the aircraft always wanted to 'weathercock' into the wind. Also, once across, high wind gusts tried hard to blow it over. Assistance from people hanging on to the wingtips was essential. During blustery weather, Alf Cormack could always be relied upon to round up a few friends in the bus belonging to the Pentland Hotel and take them out to the airfield to hold down Eric's wingtips when he came in.

February was also spent familiarising a new pilot, John Nielan, on the Orkney routes. That month also saw Eric make his first landing on the island of Hoy.

On 22nd March 1936 Eric went to Hanworth to test fly a BAC Drone aircraft for Aberdeen Flying School. The Drone was basically a glider with a single motor-cycle engine mounted behind the cockpit above the wing that pushed it along. The result was a low-cost machine that Eric noted in his log-book as being 'Good fun to fly!'. It joined the Flying School along with two Gipsy Moths that were also used for joy-riding, aerobatics and air racing. Gandar Dower charged 5/- for joy-rides at his air displays while for 7/6d game spectators could get a taste of aerobatics.

Of the other miscellaneous aircraft in the hangars at Dyce, there was an Avro 504K which was one of Gandar Dower's personal aircraft. The Avro 504K first flew in 1913. At that time petrol engines wouldn't run smoothly unless they had a flywheel and this would add prohibitive weight to an aeroplane. Someone got the bright idea of making the engine the flywheel by securing the crankshaft of a radial engine to the aircraft and the propellor to the engine. Hence two good smooth engines, the Clerget and the Le Rhône were born, both of which were used in the 504K. However carburation was a problem that required two levers, one letting air into the sump and the other letting petrol in. These were used to get the mixture correct before it found its way into the cylinder head. As petrol dissolves ordinary oil, castor oil had to be used. This would spew out of the exhausts in large quantities making the aircraft and pilots easily identifiable by their distinctive smell. There were never any constipated Avro 504K pilots!

Gandar Dower had done his training on this type when in the Royal Naval Air Service and he had bought this machine originally for towing advertising banners from Shoreham aerodrome. Having laid there it eventually found its way to Dyce where it sat in a hangar as far out of sight as possible. Eventually Gandar Dower said, 'This plane has to be flown sooner or later. You're Chief Pilot, Eric. You'd better fly it.'

While Eric had not flown this type of aircraft before, he had often heard his instructor at Redwings, Frank Payne, discussing rotary aircraft so he felt fairly confident that he could handle it. The hazard however was that if too much petrol reached the engine it choked and wouldn't start running again unless the mixture was leaned off by shutting off the petrol. When the engine roared into life it was important to remember to open up again otherwise its recovery was short lived. It had no proper 'slow running' so pilots had to 'blip' their way in by periodically cutting out the engine with a 'blip switch' on top of the joystick. Eric located the two settings that gave the most power and put two pencil marks on the quadrant. Similar marks were made for cruising and slow flying so that he could fly it on a total of six pencil marks.

Eric then instructed Henry Vallance who then instructed Roland Morris who succeeded in choking the engine and hit a boundary wall during a banner-towing display. He crawled out of the upside down wreckage unhurt but the aircraft had taken its last flight. Gandar Dower commented philisophically about the fate of his Avro 504K saying that it was 'probably the best thing for it!'

Henry Vallance had joined the team that winter and Eric had his companionship on several flights during April 1936 as he familiarised him with the scheduled routes. It was Henry Vallance who, after putting in sterling service with Aberdeen Airways, served with Highland Airways and then capped it with a long career in the service of KLM Royal Dutch Airlines. Vallance's presence took some of the weight off Eric with the resumption of the summer schedules which were not without incident and again indicated the fickle nature of the weather encountered in the north-east.

On 18th July Eric encountered what he reckons is probably his worst ever flight. Eric took off from Dyce in Dragon G-ADFI and flew straight into low cloud stuffed with thunderstorms and admits to never experiencing anything like it before or since. The aircraft was being thrown all over the place. Eric descended over the Moray Firth with the idea of following the coast but he flew as low as he dare, still without seeing the water. The nearest radio station was Kirkwall but to contact them he would have to lower his trailing aerial which, in these conditions, would just short circuit the thunder clouds. Eric continues the story:—

'All aircraft had fixed aerials running from the wingtips to the rudder. As we were using a frequency in the 800 meter bracket, this did not give much range and was only used close to the airfield. In the flight deck was a reel containing 100 feet of wire with a lead weight at the end. This was the 'trailing aerial' and it gave adequate range. However there was no future in trailing 100 feet of copper wire through a thunderstorm. One just short circuited the storm. Therefore I had to keep my trailing aerial wound in so my radio was therefore useless.

'When I estimated that I was over the Moray Firth, I came down hoping to complete my flight under the clouds. I came as low as I dare, but couldn't see the water — there were no sensitive altimeters in those days, just an altimeter marked in 200 ft. intervals. I wasn't even able to use my aerial as a 'Ground proximeter', an old trick performed by holding the end of the trailing aerial in one hand and waiting until I thought I could feel the lead weight at the end touch the water. So up I went again and continued to bounce my way through the thunderstorms.

'For the uninitiated, thunderstorms provide the worst turbulence one is likely to meet. When I estimated I was over Thurso, a lot of the turbulence had died down and I could just hear Kirkwall on my fixed aerial so I knew I was somewhere in the right area. Desperate situations require desperate measures. I let out my aerial and asked Kirkwall to give me a bearing as quickly as it could. The moment I got it I frantically wound in my aerial and headed for Kirkwall.

'Later on our people at Thurso told me that they heard me pass over so my dead reckoning must have been good. Shortly after I got the bearing I broke into the clear above low cloud. I then saw a small break in the cloud showing a bit of coast with a lighthouse on it. I immediately recognised it as one on the south coast of the island of Hoy. Down I went through this hole and fortunately there was enough space between the base of the cloud and the ground and I made Stromness safely. I still look on this as one of the worst flights I have had.'

Upon arrival at Stromness Eric went into the cabin to apologise to the passengers for giving them such an incredibly rough trip. Among their number was a dear white-haired old lady on her first flight and who must have experienced an extreme amount of discomfort. Yet, with a gleam in her eye, she responded, 'So that was a rough trip was it? That's most interesting. I've never flown before and I had my eyes shut most of the time. So that was a bad trip. Most interesting!' She had such courage, that if Eric had flown her upside down all the way, he was sure that she would have been quite happy.

(It seems that little old ladies require minimal adjustment to the concept of air travel. An example was experienced by Ernie Smith, a Chief Steward with B.O.A.C. in the early post-war period. Having been informed that a very old lady was on board making her first flight, he thoughtfully sat beside her and gently took her hand until airborne and the noise had diminished. 'I hope you don't mind?', he enquired. 'Not a bit,' she replied. 'I'm used to it. I've lots of grandchildren and they always run to me and take my hand when they are frightened.')

It has been suggested that it is strange that the frail passenger aircraft of the thirties didn't carry parachutes. However, passengers would often have been oblivious to real times of danger as opposed to times when they thought they were in danger. On a very rough trip passengers could be terrified but there was no danger, just a lot of unpleasant sensations. Yet if the aircraft was flying in complete calm above clouds that were preventing location of a landing site the passengers would not be persuaded to jump with their parachutes because they would feel so secure.

Eric recalls one trip from Edinburgh to Dyce when he was having to crawl along the cliffs of the coast with the cloud resting on the clifftops. Flying below cliff level, Eric was terrified that he wouldn't make his destination. When they reached Aberdeen the passengers came and thanked him for a most enjoyable scenic flight, totally unaware of the potential dangers that Eric had been battling against all the way north.

Highland Airways' pilot, flying from Aberdeen (Kintore) to Orkney, was Adam Smith, another experienced aviator. As the competition hotted up both Adam and Eric were reluctant to let the weather ground one in case the other went ahead with their service and got through. They both realised that this couldn't go on forever so if the weather was doubtful they would come to a joint decision over the telephone and then retire to a pub in Aberdeen for the rest of the day.

Pilots enjoyed a high degree of passenger loyalty so that regular travellers were often placing their faith in a pilot with whom they felt comfortable rather than a particular airline. If a regular passenger was wooed away from his usual pilot it generally resulted in a bit of leg-pulling. Adam Smith acquired one of Eric's regulars and jested how Eric had scared the wits out of him on his last trip by flying so high that the boats were only this size, holding his hands about a foot apart.

As the summer approached, both Highland Airways and Aberdeen Airways made plans to commence services to Shetland. The only suitable place located for an airfield was at Sumburgh which Fresson had proceeded to develop. However only the installation of radio facilities would permit its safe operation for regular scheduled services because of the lengthy stretch of sea to be crossed. Plans were now at an advanced stage for the installation of these which would of course mean that Sumburgh would have to be made available to other airlines.

As these developments were taking place, both airlines engaged in stimulating interest in the air services that they were to launch, both claiming that they were to be the inaugurators of the first Shetland services.

On 30th May Eric flew Gandar Dower and Caroline Brunning to Sumburgh via Thurso and Stromness in Dragon G-ADFI. Leaving them on Shetland to finalise arrangements for the Aberdeen Airways service, Eric then flew non-stop back to Dyce. The direct homeward flight took him two hours. He then returned on 1st June to collect them.

On 2nd June 1936 Eric inaugurated the first scheduled service to Shetland as Gandar Dower pulled a fast move to upstage his rival.

The party on board included Viscount Arbuthnott, a reporter representing the Daily Express, and the vice-president of the Aberdeen Rotary Club. The aircraft was Dragon Rapide G-ADDE, 'The Aberdonian'. They departed from Dyce at 11.00am and 1 hour 35 minutes later they touched down at Sumburgh and taxied up to the wooden building that Fresson had erected for the benefit of his passengers.

Fresson had previously advised Gandar Dower that Sumburgh wouldn't be open to him until 3rd June when the Highland Airways Shetland service would commence. As it was, Fresson was caught totally unawares.

In his memoirs Edmund Fresson reflected, 'In May 1936, we advertised the commencement of our Shetland run on 3rd June. A few days before that date an advertisement appeared in the Aberdeen and Shetland press announcing the opening of a similar service by our competitors the day before we were advertised to start. I immediately wrote to Aberdeen Airways, Dyce, a letter pointing out that Sumburgh airfield was not open until 3rd June, and refusing them permission to land until the airfield was officially opened. Foolishly, I did not send Jim Black our Sumburgh agent instructions to immobilise the airfield by placing obstacles over the landing area. I don't think, at that date, we had fully appreciated the type of forceful competition we were up against.'

At the time Gandar Dower tried to play down the rivalry with Fresson in his address to the civic authorities. One of the advantages brought to Shetland for the first time on that day was the delivery of the daily newspapers — on the day of publication! Eric, who was later to fly for Fresson, recalled the incident many years later as being 'a bit naughty'.

On the return trip Gandar Dower produced a small suitcase which contained two glasses and a half bottle of champagne so they toasted the success of the airline. It is the only time that Eric ever permitted himself alcohol when flying, during his career as a line pilot.

On the evening flight back to Dyce Eric's inaugural guest passengers were joined by the first two 'paying' passengers. This time the flight back only took 1 hour and 25 minutes.

The Shetland service often reverted to a service by the slower Dragon Moth aircraft but the Rapide was to play an increasing role in the Aberdeen Airways wider operation. When the first Rapides arrived, they had no flaps. As the fields were very short, it was necessary to approach as slowly as possible so as to land as quickly as possible after passing over the hedge. The long thin wings stalled easily. (When flaps were fitted these would be put down to make the landing approach much steeper making speed less critical.)

If a wing dropped near the ground and the pilot corrected with aileron, the extra incidence caused by the dropped aileron was enough to stall the wingtip and the wing dropped even further. All of the early de Havilland aircraft had powerful rudders and it was soon found that a jab of opposite rudder accelerated that wing and up it came. Eric got into the habit of using rudder rather than ailerons for even starting turns.

Eric's deft handling of the Rapide produced a reaction from Gandar Dower which Eric recalls, 'This creeping over the boundary as near the stall as one dare, worried Gandar Dower and he became a very bad 'backseat driver'. 'You're too high' — 'You're too low' — 'You're too fast' — 'You're too slow' and so on. I soon found a way to stop him. 'I bet you five bob I'm just right'. No money ever changed hands but it seemed to do the trick and shut him up.'

June 1936 also saw the opening of Perth Aerodrome when on the 5th Eric flew a party, including Gandar Dower and Lord Carnegie, to the official ceremony in Dragon Rapide G-ADDE.

Eric was to return to Perth six months later for a two week period as a member of the Reserve of Air Force Officers, for which he was paid a handsome retainer of £25 a year, for his reserve training. One previous course had been undertaken at Brough, using a development of the Blackburn Bluebird, the Blackburn B.2. Trainer, but his regular civil flying was considered more than enough to compensate for these being undertaken annually.

At Perth training was undertaken in a de Havilland Tiger Moth and the twenty hours of flying included forced landing practice, aerobatics, blind flying and aerial photography. The training was conducted by civil flying schools under contract to the R.A.F. and were established as a satisfactory means of assessing aspiring R.A.F. pilots before they had taken the King's shilling. It was easier to dispense with their services this way if they proved to be unsuitable.

Perth was Eric's first introduction to the Tiger Moth and it was during a particularly cold spell of weather during January 1937. He didn't take kindly to the open cockpit in these conditions and during a cross country exercise to Montrose he kept crouched as low as possible to escape the icy current of air. During the course of this he heard a strange noise and he looked out in surprise to find another Tiger Moth flying in close formation with him. The crackle of its stub exhaust was what had drawn Eric's attention to its presence as he flew in this slow open aircraft.

Eric had already been a firsthand witness some months earlier as to where the war games of reserve training exercises might eventually be heading. Under a New York dateline, the Daily Express of 19th June 1936 carried the following story:

*Airman Flies Over the German Fleet*

*An airman whose nationality cannot be revealed flew over the North Sea and took photographs of the German fleet's manoeuvres, according to a despatch from London sent by 'Augur' to the New York Times.*

*'Augur' adds that at least one great Western power now possesses accurate information regarding the Nazi's naval preparedness and methods of warfare. — Reuter.*

The Orcadian that had gone to press a day earlier did not however beat about the bush:

*German Warships off the Orkneys*

*Two German warships were operating on Thursday forenoon between North Ronaldsay and the Fair Isle. The vessels were travelling at very high speed.*

*The vessels were observed by Pilot Starling of Aberdeen Airways Ltd., whilst flying from Stromness to Sumburgh. Pilot Starling, who described the vessels as battleships, came down to 500 feet and had a close look at them.*

The normally routine scheduled flying was frequently interrupted by air ambulance flying. Some air ambulance patients travelled on the scheduled service itself and Eric remembers two unusual cases from Stromness. One concerns a lad who had swallowed his teeth and they had become lodged in his throat. The other was a member of a shooting party who had been peppered in the face. His injuries caused concern for one of his eyes and Eric had to fly him to hospital in Aberdeen.

On one occasion Eric was persuaded to fly the single-engined Puss Moth to Shetland. Northbound he had followed the coastline via Inverness but on the return journey he decided to short circuit the journey and fly across the Moray Firth from Lybster to Portsoy. Half way across his engine started to peter out with a trawler immediately below. Eric shook the aircraft about and gave it some verbal intimidation so that it started again. He then swithered whether he should ditch in the sea beside the fishing boat or try to reach land. He decided to turn back towards Caithness and, after going through another spasm of spluttering, he made Wick where it was discovered that there was water in the fuel.

The worth of the Shetland service was highlighted in January 1937. Because of continuous gales, shipping had been deterred from entering Lerwick Harbour for a fortnight but during this time the air services had continued operation into Sumburgh with monotonous regularity. Aberdeen Airways carried the first ever mails to Shetland at this time in order to help the Post Office clear the backlog that had accumulated because of the disruption to the ferry services. One aircraft carried the mailbags from Dyce to Thurso where Eric supervised their transfer to another machine which flew the Shetland leg. His camera was to hand to record this historic event.

Eric had flown Gandar Dower and Caroline in the Puss Moth to Thurso to witness the operation. This was accomplished after a false start requiring a return to Dyce, the Puss Moth again cutting out due to water in the fuel tanks. During the second attempt they had to land in a field at Novar, between Dingwall and Invergordon, to sit out a snow storm.

The launch of services across the wide stretch of open sea between Scotland and Shetland had been an intrepid piece of pioneering by all the pilots involved, both with Aberdeen Airways and with Highland Airways. But Gandar Dower had his sights set on an extension to his operation that was even more courageous. And Eric Starling was the man who was to turn those ambitions into reality.

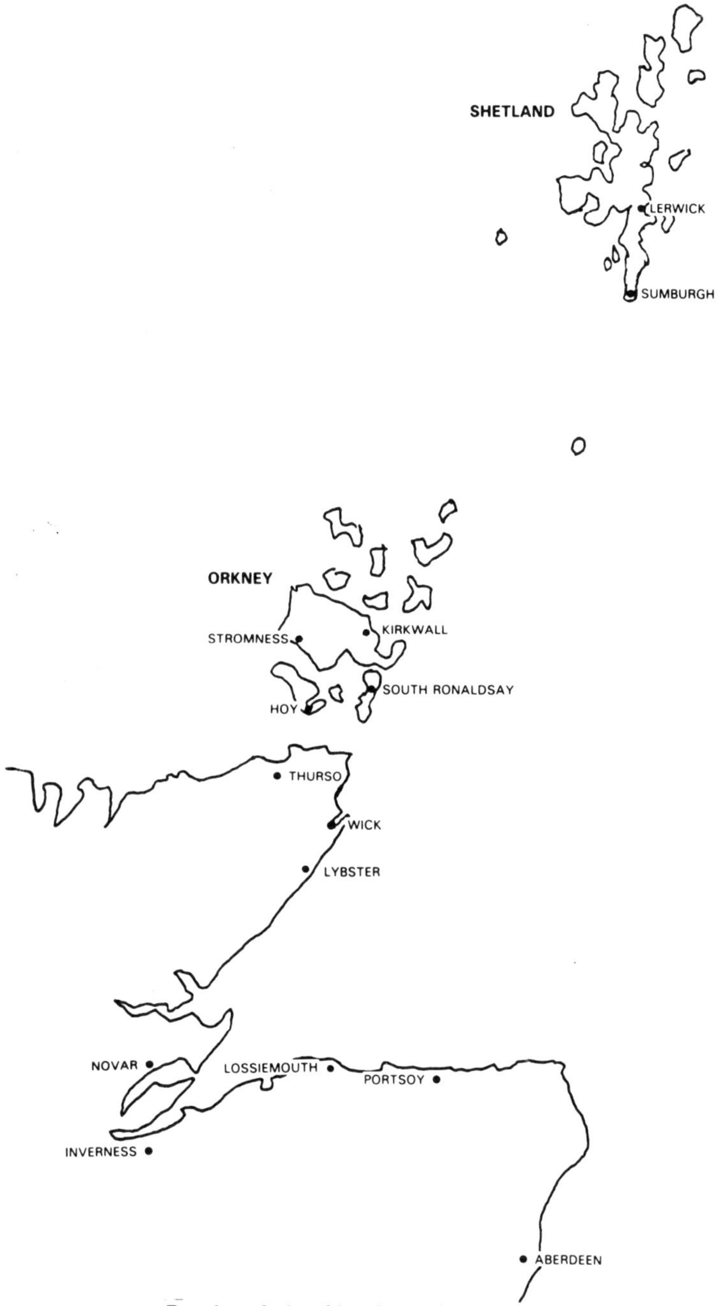


Battle of the Northern Isles 1935–36

# Norway by Airway

In 1914 an event had impressed itself firmly on Gandar Dower. At that time he was a regular contributor to 'The Aeroplane' magazine and during the course of covering features for the publication he found a unique opportunity confronting him in the north-east of Scotland.

Jens Tryggve Herman Gran was a Norwegian adventurer who had already participated in various Polar expeditions with the Norwegian Navy and had been on Scott's fatal expedition in the Antarctic. At 1.00 p.m. on 30th July 1914, after waiting ten days for suitable weather conditions, he set out on a solo flight from Cruden Bay, just north of Aberdeen, for Stavanger. His aircraft was a single engine Blériot monoplane powered by an 80 h.p. Gnome rotary engine. After being exposed to the elements in the open cockpit for four hours and eighteen minutes, at a cruising speed of 70 m.p.h., Gran completed the 293 mile journey to land at Reve, a farm located by a drained lake at Orre on Norway's Jæren coast, Stavanger, to make aviation history. The French-built machine bore the name 'Ça Flotte' but on the successful completion of the crossing it was renamed 'Nordsjøen' ('The North Sea').

Tryggve Gran's flight across the North Sea brought about a new world record at that time for consecutive flight over water. His effort was made all the more remarkable by the fact that he had only learned to fly earlier that year. Only one man, farmer Thore Reve, was on hand to welcome him as he landed 'like a huge butterfly with widespread wings', while one old lady vaulted a fence in fright. The news of his feat quickly spread through the local population so that 10,000 people gathered to cheer him on the quay when he continued from Stavanger to Oslo by the more conventional conveyance of the night steamer. Norway buzzed for days over the achievement of their hero, yet his fame did not spread across the world as it should have done, it being preoccupied with the troubles in Europe that were soon to fester into the First World War.

Gandar Dower did however take keen interest in the successful conclusion of Gran's epic flight and he did not forget this display of skill and daring. If the event had been receding from his thoughts he received a timely reminder, as his own aviation ambitions started to develop, by the reappearance of Tryggve Gran in Aberdeen in July 1934. Now a Major in the Royal Norwegian Air Force, Gran had come to Scotland to mark the twentieth anniversary of his historic flight by giving a repeat performance. This time there were three Fokker aircraft of the Royal Norwegian Air Force but the crossing was still subject to the vagaries of the weather and left a day late of the anniversary, on 1st August 1934.

Gandar Dower had played host to the hero of his youth during the Norwegian team's visit to Aberdeen. The facilities of his new airport at Dyce were put at their disposal and the airport was used as the starting point for the flight. The land of the Vikings lying across the open sea to the east was again focused clearly in the mind of Gandar Dower.

The development of aviation in Norway, dictated to a degree by its sheltered fjords but little flat land, had been based on seaplane operations. However in 1937 an airfield was opened at Stavanger and Gandar Dower was ready to take advantage of this new facility. Anticipating potential spheres of co-operation with other airlines dove-tailing their services with his own, Gandar Dower also chose this time to change the name of his airline from Aberdeen Airways to Allied Airways (Gandar Dower) Ltd and this was effected on 13th February 1937.

On 22nd May 1937 a team from Allied Airways set off from Aberdeen on board Dragon Rapide G-ADDE 'The Aberdonian' to make their own little piece of history. Accompanying Eric Gandar Dower and Caroline Brunning were Radio Officer Alec Milnes and Engineer Cecil Goodall. At the controls was Eric Starling as the airline's Chief Pilot.

In preparation for this flight Eric had already undertaken tests on newly fitted radio equipment and on the fuel consumption of the aircraft. However this first venture across the wide expanse of the North Sea was still undertaken with a degree of trepidation and anticipation. Eric would have preferred to have departed from Kirkwall which had radio direction finding equipment. Back bearings from Kirkwall would immediately let him know if he was off course. However Gander Dower insisted that tradition was followed and the flight be commenced from Aberdeen. Thus Eric only had cross bearings from Kirkwall which, while indicating his progress across the North Sea, did not indicate to him whether he was in fact heading for Stavanger. For accurate navigation over the sea precise knowledge of the wind is also necessary and this was unavailable.

In the absence of this information Eric adopted a 'known error' method of navigation. By constantly erring to one specific side of the projected route (north) he then knew with confidence that he should turn to the south to find Stavanger Airport when he hit the Norwegian coast. For this purpose Eric aimed the Rapide on a compass bearing for Utsira.

The crossing progressed smoothly into the unknown, surprising fishing boats whose crews had never seen an aircraft so far out to sea, at a speed of 137 m.p.h. and at an altitude of 1,000 feet. The weather was fine until the Norwegian coast was within reach but then it became hazy. However two islands could be seen protruding through the mist and one of these proved to be Utsira.

Eric had never previously seen Norway and the landfall that confronted him was of a jagged coastline. However he was now not far from Stavanger and made a safe landing at this new airfield. This was the first international flight to make an arrival at Stavanger and Eric found the dubious honour of being registered as 'Customs Entry No.1' being bestowed upon him.

The flight had lasted two hours and twenty-five minutes. With fuel for a flight of four hours and ten minutes they had obviously passed a point of 'no return' above the North Sea and would have been in serious trouble if they had been unable to find Stavanger Airport or if it had been fog bound. The only other airfield was across the mountains in Oslo and the terrain did not look generous in its provision of emergency landing sites. The successful and almost uneventful flight had nonetheless not been without risk.

On the following day the party continued to Kjeller military airfield at Oslo. While Gandar Dower went off to attend to negotiations and pay courtesy calls, Eric spent three days gaining first hand experience of the operations of D.N.L., the Norwegian airline.

Their services were based on the German built Junkers Ju-52. These three-engined monoplanes were fitted with floats for amphibious operations. On 24th May Eric flew Oslo-Stavanger-Kristiansand-Arendal-Oslo as an observer with Captain Finn Lambrechts. Then on 26th May he acted as co-pilot with Captain Bernt Balchen flying Oslo-Arendal-Kristiansand-Stavanger-Haugesund-Bergen, returning to Oslo over the same route the following day. Captain Balchen had already shot to fame as the first pilot to fly over the South Pole, in a Ford Trimotor in the company of Admiral Bird.

The familiar Rapide then took the party back to Stavanger for a further six days which included a fly past at the official opening ceremony of Stavanger Airport on 29th May 1937. For this operation Eric's Rapide played second fiddle to a giant Junkers G38 with four engines and with passenger accommodation in the wings as well as the fuselage.

The opening ceremony was performed by Norway's King Haakon VII. It was said that Eric bore a striking resemblance to the Norwegian monarch and this is confirmed by contemporary photographs. Eric was later told by Tryggve Gran, who was friendly with the Royal Family, that the Queen had said to King Haakon, 'If he looks like you I am definitely not flying with him!'

When Eric had made his first landing at Stavanger on 22nd May, the officials there had asked him for his comments on their new airfield. In response Eric had remarked that when he started operating the proposed scheduled service he would require them to have direction finding equipment. If at any time he had serious engine trouble he would need Milnes to act as both radio officer and cabin attendant and that wouldn't leave him with much time for the careful taking of radio bearings with the hand operated direction finding loop fitted to the DH86.

The observation had been made without great seriousness since Eric and his colleagues had become quite used to making do without adequate communications. It therefore came as some surprise to be greeted by the Commandant of Stavanger Airport offering profuse apologies that direction finding equipment had not been made ready in time for their return flight to Aberdeen on 2nd June 1937.

A strong effort had been made to rectify this short-coming of the new airport in the minimum of time. To do this they had gone to Stavanger harbour, removed a direction finder from a ship that was laid up and returned to the airport with it. 'And there it is,' he said, 'but the carpenter hasn't arrived to cut a hole in the roof for it!.'

With the addition of two Norwegian journalists the payload for the return flight was increased to seven people. The Rapide climbed to 600 feet below an overcast sky and set off into a 20 m.p.h. headwind. Visibility gradually improved as the cloud dispersed and when Gandar Dower read the boats' registration letters of a fishing fleet identifying them as originating from Peterhead, they knew that they were nearing home. This return leg of the first North Sea adventure was concluded when the aircraft touched down at Dyce exactly three hours after lift-off from Stavanger.

On the following day Eric was back on his familiar routine by flying a Dragon on the Aberdeen-Thurso run, but plans were soon well underway to make a reality of a Norway service. On 28th June Eric went to Hatfield to be tested on a new aircraft that had been specifically acquired for the Norway service. This was the de Havilland DH86B Air Express, a four-engined bi-plane which carried a crew of two plus ten passengers. It had a cruising speed of 140 m.p.h., a range of 800 miles, and important luxuries included de-icing equipment, a toilet and cabin heating. The Allied Airways Air Express, G-AETM, was named 'The Norseman'.

At Hatfield Eric spent an hour doing the usual familiarisation routines of light and heavy landings before flying 'The Norseman' to Aberdeen that afternoon. The next day, 30th June, the new aircraft was being shown off to local officials and notables with Eric carrying 48 passengers during seven short joy-riding flights.

How different from his conversion to the Vickers Viscount twenty years later with British European Airways. Firstly there was a two week technical course followed by a week on the simulator so that it was only after three weeks of training that he was even permitted to touch the real aircraft. Finally, when he had passed the necessary flying tests — coping with engine failures, flap failures, emergency lowering of the undercarriage and so on — he still had to do twenty hours flying on the routes under the supervision of a training captain before he was 'let loose' on his own.

Initial plans to launch the Stavanger service from Aberdeen were changed with the substitution of Newcastle. There were several reasons for this choice and these included the availability of good radio facilities, connecting flights from London by North Eastern Airways, and Newcastle's established position as port of embarkation for sailings to Norway.

It was therefore from Newcastle that Eric set off in 'The Norseman' for another proving flight to Stavanger on 5th July 1937. On this occasion the party of seven continued to Copenhagen and Gothenburg, returning to Newcastle's Woolsington Airport on 7th July.

The inaugural flight, the first scheduled flight between the United Kingdom and Norway took place on 12th July 1937. Eric was accompanied by Milnes, his wireless operator, and six passengers on the outbound flight, and on the return leg there were eight passengers, the Mayor of Stavanger and his aviation committee going to meet their counterparts in Newcastle.

On this return flight the aircraft became tail-heavy as they approached Newcastle. Eric sent Milnes back to see what the problem was and he discovered all the passengers tugging at the toilet door. The Mayor of Stavanger was locked in the lavatory! On arrival at Newcastle he expressed his relief that there were no members of the Press on board.

On 10th August 1937 Eric flew Major Tryggve Gran to Newcastle to see Gandar Dower. Major Gran was accompanied by his daughter whom Eric remembers vividly as being a beautiful Norwegian blond. However Gran, in the frequent compliments paid to his daughter, repeatedly added 'and she's only fifteen'. Eric got the message!

The route soon settled into a regular pattern, although passengers were a scarce commodity being slow to take up the company's slogan, 'Norway by Airway'. On one occasion Newcastle was fogbound and Eric had to divert the returning flight to Carlisle while Aberdeen was used a diversionery field a couple of times. On another occasion Eric was crossing the North Sea in fine weather conditions when ahead he could see a solid wall of thunderstorms. Deciding that discretion was the better part of valour Eric executed a gentle U-turn and headed back to Newcastle.

Gandar Dower was in Newcastle that day and was accosted by an angry City Councillor demanding an explanation for the abandonment of the flight 'on a fine day like this'. Gandar Dower had been unaware that Eric had returned but promptly replied that 'the Directors of Allied Airways are perfectly satisfied'. He did not pursue an explanation from Eric until he volunteered it and that increased Eric's esteem for his employer.

On another occasion Eric was proceeding on a nice clear day when Stavanger came on the radio saying that they were fogbound and that Eric couldn't land. A few days later a young lady passenger enquired if they would get to Norway that day. Eric asked of her why she thought that they shouldn't. 'My father is a sea captain and he saw you turn round and come back just the other day!' was her reply. Eric had noticed a few ships below him that day but never suspected that his movements would be reported back by one of their captains.

On other days the flight across the North Sea was so calm and uneventful that Eric could sit and read a book while at the controls of his aircraft. He didn't have an auto-pilot of course, but if he had no passengers or if the occasional passenger was snoozing, it was very easy to keep the aircraft in trim and read with just a cursory glance at frequent intervals across the empty air space above the sea.

On 27th September, after forty-four return crossings, the Newcastle-Stavanger service was withdrawn for the winter. A feature of Newcastle's grass airstrip had been its fine mushrooms which Eric carried off in disposable air-sickness bags. Gandar Dower had other diversions planned for Eric through the 'quiet' winter months.

Having awarded the contract for the first U.K. internal air mail service, between Inverness and Kirkwall, to Highland Airways in 1934, the Post Office were sufficiently impressed by it to wish to operate a similar service to Shetland. Gandar Dower tendered to carry mail from Dyce to Shetland and 'won' the contract.

This was to be Eric's nice little job for the winter. Edmund Fresson of Highland Airways did not tender for the Shetland mail contract and reputedly laughed his head off when he heard that his rival had taken it on, laying bets that the Allied Airways mail service would become a farce through cancellations. Certainly with the absence of radio and navigational facilities at Dyce even Gandar Dower obviously had some misgivings and consoled Eric with the news that he had already warned the Post Office that they would probably have to cancel flights quite often.

Eric flew a newly acquired de Havilland DH84 Dragon from Hatfield on 19th November 1937, arriving at Dyce two days later after calling at Sywell where £10 was made by giving joy-rides for Northampton Town Council, followed by stops at Doncaster and Newcastle. This aircraft, G-ACNJ, was named 'Sir Rowland' after the founder of the Post Office, Sir Rowland Hill.

After due ceremony on the morning of 23rd November 1937, during which the Royal Mail pennant was presented to Gandar Dower, Eric set off in 'Sir Rowland' to fly the first Air Mail service to Shetland. Mail weighing 480lbs was carried via Stromness to Sumburgh, and 627lbs of mail was carried on the return flight which called at Stromness and Thurso. The innovation of the air mail was not only to knock some twenty hours off the transit time by sea for letters but was also several hours faster than the telegram service. At that time the telephone had yet to reach the Shetland Isles.

To the surprise of everyone, including Eric, the service operated right through until March 1938 before the weather was to force the cancellation of a service. Eric had discovered that regular flying in the winter was easier than in the summer provided the gales could be coped with. In winter the weather changed so quickly that an airfield that was fogbound one minute was cleared the next. During summer the fog could lie for two or three days.

During the pre-war era no official meteorological services were provided. Pilots used to ring up the coast guards and so build up their own 'weather lore'. For instance, they knew that they would always be able to land at Wick if the wind was offshore, while Inverness would outrival Prestwick for good flying weather. North-east winds would occasionally bring in fog at Inverness but that seldom happened.

Shetland did not present any suitable diversionary landing sites if Sumburgh was fogbound so the only option was to turn back. However Eric and his colleagues still seemed to enjoy a casual approach to the now familiar Shetland airfield. Eric would contact Jim Black at

Sumburgh for a local weather report. He would calmly reply, 'You'll get in alright. You'll get in. We can just about see Sumburgh Lighthouse.'

On 8th April 1938 the Aberdeen Press and Journal reported, *'March was an exceptionally bad month in Shetland. It was the windiest and coldest on record, the wettest for twelve years and the most sunless for ten years. The highest temperature, on any day, of 53 degrees was lower than Aberdeen's average maximum for the month of 54.7'* Eric had coped with it all except on one day when visibility was down to 200 yards (with the lighthouse presumably totally obscured) and after circling for thirty minutes he had to fly to Stromness.

On the first anniversary of the service the newspaper was applauding the success of the Shetland air mail, *'Wonderful regularity has been maintained during the first year of service of Allied Airways Dyce-Shetland Royal Mail 'plane, and during that period no less than forty-four tons of mails were carried. The failures in regularity of outward and inward journeys have been few indeed, and Allied Airways may be proud of this achievement. On two occasions only could the 'plane not leave Dyce until the following day. Tributes have been paid by the Post Office to Allied Airways for the dependability of the service.'*

However Eric's role in the Shetland air mail had taken a break by April when he returned to Newcastle to restart the Stavanger service for its second season on 16th April 1938. The flights at the controls of 'The Norseman' now settled to a regular pattern. On 2nd May Eric recorded a perfect day for his 100th crossing of the North Sea.

These flights would be punctuated by Army Co-op flights, night flights to give the army searchlight practice. These would sometimes be flown in a Puss Moth but the DH86 Air Express was also used in this role. Joy-riding flights also appeared from time to time and these five minute flips for fun-seekers must have been a sharp contrast to the monotonous hours crossing the North Sea. Eric also occasionally took the DH86 Air Express north to operate on the Orkney run where its appearance must have been a novelty for travellers used to the Dragon aircraft.

On 19th September Eric made his 166th crossing of the North Sea in 'The Norseman'. This marked the end of the 1938 summer season for Norway and Eric returned to Aberdeen. The service had not enjoyed the level of support required to prevent it running up a substantial loss and as it happened it was not to operate again. 'The Norseman' suffered an ignominous departure some time later when it was hastily sold off to pay out a claim by one of the airline's own directors. The aircraft was subsequently destroyed on the ground at Malmo on 2nd May 1940 while in the service of the Finnish Naval Coastguard Service.

Throughout the two years and 166 North Sea crossings which Eric had made between Newcastle and Norway, he had been safely guided into Stavanger Airport by the direction finding equipment that had been acquired from the ship in Stavanger harbour. The hastily requisitioned equipment had taken on a permanent role at the airport it seems. However facilities that had been provided at Stavanger's new Sola Airport at great expense, included two concrete runways, among the first at any European airport. Eric nonetheless insisted on irritating the airport authorities by landing on the grass which produced much less wear and tear on the landing wheels of the Air Express.

Upon his return to Dyce, Eric resumed flying the familiar northern routes once more as the winter of 1938/39 approached. But his dynamic career as Chief Pilot was now drawing to a close. Since joining the airline it had been a six day a week job. 'We never had time off and we never wanted it.' But when Eric married Eleanor Castel at St Machar's Cathedral, Aberdeen on 14th January 1939 he used the occasion as a suitable excuse to leave Allied Airways.

Eric and Eleanor had first met at Dyce Airport but it was at the annual grand ball held at the Aberdeen Beach Dance Hall that they established a rapport. Eric had been invited to a party there by Dr Agnes Thompson, a student of the Aberdeen Flying School and a close friend of Eleanor. Eleanor was always at the ball to write it up for the Aberdeen Press & Journal and she would join the party upon completion of her press assignment.

Eleanor was a fine horsewoman and Eric became a reluctant enthusiast of this sport. Sunday mornings were then frequently spent on horseback followed by lunch at some hotel. On Monday evenings Eric would be Eleanor's guest at His Majesty's Theatre to which she would be given complimentary press tickets. Eleanor would take notes on a 'Players' cigarette packet. After the performance they would rush to the office where she would write her review for the morning paper and then they would go off to enjoy themselves for what remained of the evening.

Eleanor wrote a regular column called 'A Country Diary' for the Press & Journal under the pen name of 'The Magpie'. She also wrote children's stories and plays for the paper and for the B.B.C. These are still being enjoyed by children today. Long before meeting Eric she had flown on Fresson's Highland Airways and reported on her visit to Orkney. She had more recently reported on a similar journey with Eric when the destination had been Norway. A horse made of wooden blocks that sat above the control panel of 'The Norseman' was a mascot that had been a gift from Eleanor.

Dr Agnes Thomson, Aberdeen's first woman doctor who had unwittingly been responsible for bringing Eric and Eleanor together, toasted the couple at a party held to mark their marriage. Despite a couple of flattering inaccuracies, it was a fitting send-off for the newly-weds.

'The day I took Eleanor out to Dyce and left her there to get 'copy', I did not know that I had started an event which would lead to this romantic wedding today. I was called away and left her in the charge of Eric. They had that similarity of character which at once attracted them to each other — honesty, humour and a great zest for life — they also had many tastes in common, especially their interest in horses.

'They also both possessed to a marked degree the qualities of courage, versatility and the desire always to give of their best — qualities which, in their different ways are as applicable to a career in journalism as in flying. We admire them in this strength of character, and the perseverence which has taken them both to high and responsible positions.

'I remember the morning at Dyce when Eric set off to fly the first flight to Norway in a Rapide, and as I pinned a spray of white heather to his coat and wished him the best of luck, I admired the pluck he was showing in making this flight. He didn't know if he had enough petrol to get him there! But still his resolve won through, to win for him the renown of being the first pilot to make the trip from Scotland.

'So this marriage starts off under the very best auspices, and I am sure that when the dark or difficult days come, testing times which lie ahead (as they do in all marriages), Eleanor and Eric have the spirit and loyalty which will carry them through. I know they have the pluck and wit and so will make the best of it.

'Eleanor has had to make the difficult decision of giving up a career in which she has shown much talent, but like many women before her she counts the world 'well lost for love'. Love is the greatest and most beautiful force in nature, the source of all inspiration and generosity of the spirit.

'We shall miss them both. We shall miss Eric at Dyce, and regret that he has to leave us for flying duties in England. We shall miss Eleanor's articles and gossip columns which we always read with much interest, her joyous nature and the radiance of her personality.'

The Best Man on the occasion of Eric's wedding was his close friend and fellow Allied Airways pilot Harry Reed. Harry was later to become Allied Airways' Chief Pilot but his career was to be cut short on 21st May 1941 when he had to wrestle his aircraft onto the clifftops of Caithness after it had inadvertedly been sent out from Thurso without being refuelled. The passengers on board escaped injury but the impact on landing into a stone wall was to make Harry Allied Airways' sole fatality in grappling so successfully with such a dire emergency. Sadly his shattered aircraft was Dragon G-ACAN, 'The Starling'.

When Eric flew his last Allied Airways service on 9th January 1939 from Dyce to Sumburgh with calls at Thurso and Kirkwall he had flown 3,110 hours of which over 2,500 had been for Allied Airways. His last flight was at the controls of de Havilland Rapide G-ADAH, 'The Thurso Venturer'. This aircraft was later renamed 'Pioneer' and this is the name that it still carried in retirement at the Museum of Flight, East Fortune, Lothian.

The next port of call on Eric's flightpath was to be Brooklands Aviation at Sywell in Northamptonshire.

Norway 1937/38

# The Lull before the Storm

Straight after their marriage, Eric and Eleanor moved to Sywell near Northampton, where Eric took up a position with Brooklands Aviation as a navigation instructor. He had already trained in this role on a course at Broxbourne in July 1934 in preparation for his job at Aberdeen where Gandar Dower's activities included Aberdeen Flying School.

During the thirties Royal Air Force pilots and navigators were trained at civil flying schools. This system allowed the R.A.F. to sort out the wheat from the chaff among flying hopefuls before assuming full responsibility for them. Of the various flying schools, those operated by Brooklands Aviation enjoyed the highest esteem and at Sywell they had a substantial fleet of training aircraft. Pilots were trained at the Elementary Flying Training School which had some twenty to thirty Tiger Moths while the Civil Air Navigation School, which Eric had joined, had ten R.A.F. twin engined Avro Ansons.

Brooklands paid their instructors well and this attracted several pilots with already well-established reputations. Wing Commander Grieve, who had put Eric through his instructors course four years previously, was now with Brooklands as was Ken Waller who had made history with his record-breaking flight from England to New Zealand in the original Comet, the de Havilland DH88. Eric and Eleanor settled down to what Eric describes as 'a very pleasant life which provided us with a bungalow by the airport with a daily maid, no instruction at weekends with additional days off in between courses and a handsome salary of £750 a year'. The bungalow had previously been the home of Tommy Rose, a famous long distance aviator. Next door to the bungalow were kennels where the proprietors bred Bloodhounds and Bedlingtons. Eleanor was always threatening to buy one of their Bloodhounds. It was her plan that she would use this to seek Eric out when he was drinking with the boys and she didn't know in which pub to find him.

This contrasted strongly with the irregular operations with Gandar Dower which meant service six days per week and frequent tours of duty at different outstations. The luxury of holidays and additional leisure time allowed for visits to various neighbouring parts of the countryside and the newly weds were enjoying life to the full. However it was becoming increasingly apparent that war was not far away.

Eric reported for duty at Sywell on 16th January 1939, just two days after his wedding, to give instruction at No. 9 Air Observers Navigation School (A.O.N.S.). His first job was to join some fellow new recruits to collect Avro Ansons from A.V. Roe's factory at Woodford, near Manchester. It proved to be an easy aircraft to fly, but on that first flight none of the pilots were tempted to meddle with a feature that was new to them all — a retractable undercarriage. The wheels of the aircraft, which required to be cranked up by hand, remained lowered throughout the fifty-five minute flight back to Sywell.

The trainee navigators were taken aloft during morning and afternoon sessions. Normally two trainees would be carried and they would work in turns at a map table behind the flight deck. When they weren't working with pencil and protractors they got the laborious job of cranking the wheels up or down, the pilots not only having learned to make use of the facility but also that it was a job for the 'passengers'.

Some trainee navigators failed to qualify, not through any short-comings in their mathematical skills, but because of an inability to cope with the motion of the aircraft. They were not the only casualties of air sickness. One pilot, when they later moved to Blackpool, had to be turned down because of repeated air sickness. When questioned as to when this problem had started he asked if he could make reference to his flying log. He then replied with a precise date. Wing Commander Grieve asked why it had started on this particular day. 'That was when we flew under the high tension cables at Fleetwood', was the response.

Another incident that would have understandably upset a lesser pilot's attitude to flying confronted the Brooklands second-in-command, Goldsmith. On a hazy day with low cloud, he wanted to practise steep turns with a pupil. One of the other pilots suggested that he climb above the clouds which topped dead level at 2,000 feet. This was ideal for practising steep turns so he flew away from Sywell then commenced his climb through the clouds. Suddenly the wheels of another Tiger Moth appeared above him. They crunched into his centre section, bounced off and disappeared again into the cloud. Both aircraft landed safely, the only damage being a badly tattered section on Goldsmith's Tiger Moth. Yet if the collision had involved the climbing and descending aircraft making contact at almost any other point, the freak meeting would have resulted in the most serious consequences.

Because of the bleak future that seemed to be confronting all of Europe, Eric and Eleanor had decided to delay any plans of starting a family. However the young couple must quickly have settled into the social scene at Brooklands because Eric was soon reporting 'a change of attitude after a good boozy party at the airport one evening when Eleanor became pregnant'. Their worst fears for the days that lay ahead were confirmed on 1st September when Germany invaded Poland followed two days later by the British declaration of war on Hitler's regime.

In December 1939 Eric was moved to Blackpool. Just before this the relationship with the R.A.F. was formalised. Eric and his colleagues were put into uniform and drafted into Coastal Command as members of No. 2 School of General Reconnaisance. Eric became a senior Flying Officer as a result of his participation in the Reserve of Air Force Officers during the previous five years. Some other instructors with equally fine flying pedigrees were only Sergeant Pilots having been Volunteer Reserves.

The change was taken with tongue in cheek since they all continued with the same duties while also 'playing soldiers'. The ranks frequently didn't conform to the pecking order within Brooklands Aviation. Duncan Davies, the chief and founder, was a Flight Lieutenant while one of the part-time weekend instructors appeared as a Squadron Leader. Goldsmith, the Chief Instructor, was a Flight Sergeant with dozens of officers working under him.

Few of the instructors pressed into uniform had ever sat the official R.A.F. Wings examination. Wingless flying instructors didn't look right so one of the team made a visit to Bedford to purchase thirty sets of R.A.F. Wings to sew on to the uniforms. At Sywell there were no Messes and the instructors adjourned to the clubhouse bar for lunchtime beers. When a visit was paid to Sywell by Sir John Steel, Chief of R.A.F. Training Command, he was aghast to witness officers and sergeants cheerfully drinking together. He disapproved of beers between flights almost as much as he frowned upon this mixed company so he wasn't put any more at ease when Ken Waller, a distinguished aviator but ranked as a mere Sergeant Pilot, addressed the Air Chief Marshall innocently with, 'What about a beer for yourself, old boy?'

Tony Spooner, a close friend of Eric's at Brooklands and ranked as a Sergeant Pilot, was summoned before 'Wing Commander' Grieve for some misdemeanour perpetrated in one of the school's aircraft. After Grieve had given him a telling off, Tony asked if he would repeat himself. 'I couldn't hear what you were saying, Sir. I've just landed and I'm still a bit deaf.' Grieve repeated his reprimand a couple more times, but when Tony said he couldn't hear what the Wing Commander was saying and offered an ear at closer quarters for Grieve to have another go, his superior gave up and dismissed him. Such was the view of authority from both perspectives.

It was also Tony Spooner who endeavoured to persuade the medical officer to remove his appendix. He was in perfect health but reasoned that the odds were that they would have to come out at some stage of his life. Why not have them removed just now while it would cause little inconvenience and could be done at the R.A.F.'s expense? He was very annoyed when the medical officer didn't concur with his request.

Philip Pashley was another member of the team and he had a long aviation pedigree. He had been an engineer at Shoreham when Louis Bleriot had crossed the Channel in 1909. Pashley claimed to have flown Bleriot's aircraft, as he was responsible for repairing it and he always maintained that it had become airborne while he was moving it from one side of the aerodrome to the other. In 1911 he had been persuaded by a youthful Eric Gandar Dower to take him aloft for his first flight. Pashley could hardly have imagined the consequences of that particular introduction to the air.

Philip Pashley had served at Shoreham rigidly for thirty years, even during call-up between 1914 and 1918 when he remained there in the army as a test pilot. He was therefore quite put out when ordered to Sywell as a Sergeant Pilot, and to add insult to injury he was refused entry to the cocktail bar during a visit to Bedford as he was not an officer. This insult was inflicted on the same man who had just recently been honoured at Shoreham with a dinner to mark his completion of 10,000 hours flying instruction, a considerable feat. Perhaps he had the last laugh when he was later commissioned and posted to South Africa. A young Air Officer Commanding thought it undignified that this now elderly pilot should be slaving away teaching pilots and he sent several messages to London requesting permission to ground Pashley. Eventually a response was elicited, 'Pilot Officer

Pashley will himself decide when he wants to stop flying!' When Philip Pashley did finally decide to give up flying instruction he was aged 72 and had enjoyed a flying career which had spanned 52 years.

The doctors in uniform playing at life in the armed services in the modern television series 'MASH' remind Eric of the atmosphere at the training school at Sywell and Blackpool in 1939/40.

Eleanor preceded Eric to Blackpool to organise a house and a nursing home for the impending birth of the baby. She was accompanied by Dorothy Peacock, the fiancee of Ronald Bliss, one of the other navigation instructors from Sywell. Dorothy viewed the very pregnant Eleanor nervously throughout the car journey northwards, seeing her as a travelling time-bomb.

Eric and Eleanor moved into a furnished house in Squire's Gate Lane, close to the airfield. Their son Angus was born in Blackpool on 18th December 1939.

The airfield was at the other side of the road and Angus nearly came to an untimely end in the garden. There was a flight of Fairey Battles drogue-towing for gunnery practice. The drogue resembled a windsock and aircraft would fire at it with bullets that had their tips painted. The paint would mark the holes made by the bullets in the drogue and this aided identification of the aircraft who had made successful hits.

The Fairley Battle would release the drogue and its towing cable before coming in to land. On this occasion the drogue had been shot off just leaving a large lead weight at the end. The pilot thought he had dropped it on the airfield but it had become caught up. The pilot came in over the Starling house to land. Angus was in his pram on the front lawn at the time and the lead weight whistled through the garden only just missing it. After that more care was exercised in choosing a place for the pram when the wind was blowing from the south.

The instructors arrived here as Brooklands staff but very soon the R.A.F. took over and initially they came under a Coastal Command Squadron as No. 9 Air Observer Navigation School.

It was decided that the navigation instructors should know something about sea navigation as their work up until now had been over land. Instruction on the subject was provided by a pimply-chinned Flight Lieutenant who expounded primitive theories on triangles and velocities.

The class of seasoned aviators laughed the unfortunate youth out of class, led by a former Imperial Airways captain nicknamed 'Bad Weather McIntosh' who had some 20,000 hours to his credit. His place was then taken by Flight Sergeant Mountjoy who had been a navigator on Sunderland Flying Boats. 'Flying over the sea is easy,' he said. 'You steer in a straight

'For E Allen Starling from Ian Hamilton in memory of a record and dashing descent by an Inspecting General upon a Parade in a Puss Moth on 15.7.35'. The General's inscribed portrait commemorates Eric's impromptu landing at Nairn.

De Havilland Rapide G-ADDE, 'The Aberdonian'. Eric is in the white overalls while Captain Henry Vallance poses in sports jacket and flannels.

2nd June 1936 and the first scheduled flight has just arrived at Sumburgh Head, Shetland. Lining up before 'The Aberdonian' are Eric Gandar Dower, Eric Starling, Viscount Arbuthnott and Caroline Brunning while petrol cans for refuelling are also very much in evidence.

Dragon G-ADFI, 'The Silver Ghost', taxies out on the snow at Thurso.

Eric at the controls of Dragon G-ACAN, 'The Starling', 'somewhere above the Moray Firth'.

Harry Reed at the controls of 'The Starling' while engineer, Ben Brown, looks on from the ground. Harry was to become Allied Airways' only fatality and the tragedy occurred while at the controls of this aircraft.

Eric Gandar Dower's Avro 504K shares the hangar at Dyce with a Rapide.

At the controls of 'The Norseman'. Above the control panel is a wooden horse mascot presented to Eric by Eleanor.

line and you have a drift sight so if you are drifting 10° to port, turn 10° into the wind. All you want to know is how far you've gone and to do that you need to know the wind strength. To find the wind direction and speed you must fly a 60° triangle of velocities. Turn 60° to starboard for two minutes and take the drift,turn 120° to port for two minutes and take the drift again then continue on course. Feed this into a simple calculator and you have the correct wind speed and direction and you can find your speed over the sea.' Mountjoy quickly took the mystique out of sea navigation and this is what the instructors then proceeded to put into practice.

During the following months Eric took trainee navigators aloft for flights lasting anything up to five hours. There was the occasional visit to Prestwick but, as often the aircraft had no reason to land again until it returned to Blackpool, it would be logged vaguely as Blackpool to Ireland and return, Blackpool to Wales and return, or after mobilisation from 1st January 1940 purely as 'Air Observer Navigation Training'. In August this routine was supplemented with a couple of days of 'Bombing Practice'.

In November 1940 Officer Commanding, Training Squadron, No. 2 School of General Reconnaisance, Blackpool, assessed Eric's proficiency as a pilot as 'above average'. At that point Eric had completed 3,920 hours in solo flight. This assessment marked Eric's departure from the School of General Reconnaisance and from Blackpool.

During 1940 the Battle of the Atlantic had been hotting up and Churchill ordered that Bomber Command should give up a Squadron of Wellingtons to Coastal Command to be used for submarine spotting over the Atlantic Ocean. Bomber Command reluctantly gave up some of their aircraft while Coastal Command wondered how they would fly them with no suitably qualified pilots.

This problem was put at the door of Bomber Command who were already smarting from being deprived of their aircraft. However they eventually agreed that Coastal Command should provide four pilots with sufficient experience to require minimal training on the Wellington. Bomber Command would convert them as Wellington pilots and they would then assume responsibility for training further Coastal Command pilots.

Eric was selected as one of the 'chosen four' and on 22nd November 1940 went to Hampstead Norreys in Berkshire where Bomber Command had a Wellington Training School.

The railway network was already in a frequent state of disruption by this time and this was illustrated by the journey south from Blackpool. Eric and three fellow General Reconnaisance School instructors journeyed to the Midlands where they had to change trains but there they were told that they could proceed no further that night due to air raids in the south snarling up the system and creating a shortage of rolling stock.

A long time friend of Eric's, Tony Spooner, remembers the predicament well. 'While digesting the news and wondering what to do next, the ever curious Eric noticed an engine with a couple of carriages. The small engine had steam up and we enquired where it was going, thinking it might be proceeding further southwards. Since we had all been at Sywell, we were hoping that it might pass near Northampton where we had friends and knew a couple of friendly publicans who would, we hoped, put us up for a night.'

'To our surprise, the driver of the small train said that he would take us to Northampton, if that was where we wanted to go, but for the fact that he lived locally and that he hadn't the money to pay for a night away from home. Eric suggested that if we had a whip round and collected 22/6d, would he then take us to Northampton? The private deal was struck without any apparent concern on the part of the driver of possible head-on collisions. We interested a few other stranded passengers and were in the process of collecting the required sum when we were informed that a larger bunch of passengers had talked the driver into taking them to Leicester instead! We had been outbid! Central control of the rail system had broken down to such an extent that this kind of private *ad hoc* arrangement, almost akin to bribery, was apparently acceptable.'

At Hampstead Norreys Eric carried out a total of approximately seven hours flying, including some night flying, after which he was passed out as a proficient Wellington pilot.

During Eric's brief visit to Hampstead Norreys, a lost Hurricane arrived one night. When parked safely to the side of the flarepath, the pilot asked if he could stay for the night. 'Sure, but you'll have to wait a bit until flying is finished because we are short of transport', and he was kept parked at the end of the flare path. When the visitor eventually got to the mess he peeled off his flying gear to reveal gold braid up to his elbows. He was in charge of Fighter Command at Bristol where he was unhappy with the poor success rate of his pilots in bringing down aircraft over the city. Having listened to their excuses he had decided to take an aircraft up himself. The result was that shortly after becoming airborne he was blinded by all the searchlights and, when he had escaped their glare, he had also lost his bearings.

From Hampstead Norreys Eric went to Bircham Newton in Norfolk where he helped to form No. 221 General Reconnaisance Squadron. Here Eric trained other pilots on the Wellington and they in turn passed their newly acquired skills on to others.

Bircham Newton, situated some miles from Hunstanton and King's Lynn in Norfolk proved to be a desolate place in winter. The mess was full and many airmen occupied digs or abandoned houses. Eric and Eleanor were fortunate enough to get accommodation at a pub on the coast of The Wash at Thornham.

While here Eleanor and Angus had an abrupt reminder of the war that was going on. For a second time Angus nearly came to an untimely end when Eleanor was pushing him through the village in his pram. A German intruder flew down the main street spraying it with bullets but fortunately no one was hit.

It became evident that there was a German pilot who knew the Norfolk area very well, probably from pre-war flying. When the clouds were very low and visibility poor, he used to come in spraying villages with bullets and dropping sticks of small bombs on R.A.F. airfields. On one occasion he dropped a stick of bombs too low and instead of blowing a series of holes in the runway, they fell flat and didn't explode. The armament officer defused them while being watched from a safe distance. Having defused the last bomb, he held it above his head and threw it on the ground in a symbolic gesture of defiance.

On another occasion the flyers were having lunch in the mess on a day when there was no flying because of low cloud and mist. They were just thinking that it was exactly the kind of day for their German 'friend' when a great commotion ensued with the ground defences firing at this intruder. A few minutes later an R.A.F. Regiment officer entered the mess holding a piece of a German bomb bay door above his head. Having had his bomb bay doors shot off at Bircham Newton, the German moved on to drop his bombs at a neighbouring Wellington Station. Here fitters dashed out of a hangar, where they had been working on a Wellington up on trestles, to see what was happening and were just in time to see a bomb roll under the aircraft and explode.

At Bircham Newton Eric performed his one and only 'wheels-up landing'. Unable to lower his undercarriage, the grass field at Bircham Newton cushioned his landing to such an extent that Eric could claim a smoother landing than many he had performed with his wheels down. The Wellington had a large girder along its belly from which bombs were hung and this acted as a keel making it highly effective for such an unconventional return to the ground.

This girder was covered only with wooden fabric fairings. The girder was fixed to the aircraft with a large quantity of bolts but it was a relatively minor task to jack up the aircraft, undo the bolts and install a replacement girder and fairings. The Wellington had a weak tail wheel and if it was given a good thump on landing when heavily loaded, it collapsed and the geodetic construction at the tail end became badly damaged during the landing run. This involved a much more complicated and lengthy repair job than that from a belly landing.

As it happened, the Wellington had an emergency system for lowering its wheels which Eric tried with no result. This consisted of a hand pump which Eric had tried unsuccessfully but he had been unaware that this pumping had to be maintained for about fifteen minutes before it lowered the wheels.

Night landing practice was done at a grass satellite nearby because, in the event of night intruders, there was very little to damage. Intruders had a habit of shooting aircraft up during take-off or landing when navigation lights had to be used to let the people on the ground see what was going on. Because of this the procedure was to switch off the runway and aircraft lights immediately after take-off. After a carefully timed circuit the landing light was switched on and hopefully a flarepath lit up ahead. It was during this type of manoeuvre that Eric was involved in another minor crash. He flew with another pilot to practice night landings, completing three take-offs and landings. Then the other pilot had a go but on his first landing he came in too fast and went clean through the far hedge causing some damage to the Wellington. No one was hurt but the pilot's dignity suffered a knock.

By May 1941 Eric and his colleagues had mastered the Wellingtons, including the undercarriage mechanisms, and were ready to assume operations by way of a move to Limavady, Northern Ireland.

INVERNESS
ABERDEEN
GLASGOW
PRESTWICK
LIMAVADY
BELFAST
FLEETWOOD
BLACKPOOL
DUBLIN
MANCHESTER
WOODFORD
HUNSTANTON
THORNHAM
BIRCHAM NEWTON
KING'S LYNN
SYWELL
NORTHAMPTON
BEDFORD
BRISTOL
HAMPSTEAD
NORREYS
LONDON
SHOREHAM

# Across the Sea from Ireland

On 2nd May 1941 Eric, with a crew complement of nine, flew low over Lough Foyle to land on a newly constructed airfield at Limavady, Northern Ireland. Here he received a boisterous welcome from bystanders. This was not because Eric was recognised as some kind of folk hero honouring the station with his presence. Instead it was because Angus's pram was prominent among the luggage being unloaded from the Wellington and this was quickly the subject of a stream of witty remarks.

The Wellingtons of 221 Squadron flying from Limavady were fitted with A.S.V., equipment so secret that the meaning of the letters were only revealed with reluctance to those who were to operate it. They stood for 'Air to Surface Vessel' and this sophisticated apparatus was the first airborne radar. Antennae, similar in appearance to modern television aerials, protruded from the fuselage and wings of the Wellingtons so that they took on the appearance of spiny sea urchins.

The A.S.V. equipment provided a graphic display on a cathode ray tube as opposed to the pictorial display that accompanied later radar developments. A vertical line on the screen was marked in miles and an echo travelled down the screen as a 'blip'. The aerial protruding from the nose of the aircraft sent out radio pulses which were picked up by other aerials hanging from the ends of the wings. By timing the interval, a 'blip' was put on the cathode ray tube at a position corresponding to the distance from the object which was sending the echo. If the target was straight ahead, the 'blip' stretched equally each side of the vertical distance line. If the target was to the right, the right receiver aerials received a stronger signal and the right-hand side of the 'blip' increased in size. The aerials on the side of the aircraft had a fifty mile range and by switching over to these a one hundred mile wide track could be swept, although the path ahead was removed from observation while these side aerials were in action.

However the magic A.S.V. equipment proved to be of questionable value in detecting U-boats, the few sightings being detected by the naked eye long before the tell-tale features could be discerned from the kaleidoscope of dancing images on the cathode ray screen. This was because ocean waves also sent back blips so that a small object like the conning tower of a submarine tended to be lost in the clutter of wave echoes.

But while A.S.V. was not effective in successful U-boat detection, it did pin-point larger masses enabling the crews to find shipping convoys that they had been ordered to pick up in the Atlantic and escort to port. On one occasion Eric was detailed to escort a lone merchant vessel which was located with a degree of pride, but unfortunately one of his gunners accidentally let off a round of bullets at it. Not surprisingly the crew kept below decks while receiving the benefit of this dubious escort. A.S.V. also proved a valuable aid in letting the Wellingtons find their way back to base, giving their crews the confidence to undertake longer range sorties which would see them being airborne for up to nine or ten hours at a time.

The airfield at Limavady consisted of hangars and huts scattered over a wide area while the aircraft were also dispersed widely about the field in case of an air attack. In wet weather it was a miserable place immersed in a sea of mud which the crews had to negotiate by using old buses or bicycles.

Beyond the airfield the world seemed to be quite different from that which had been imposed on Great Britain and was supposed to be operative in Northern Ireland also.

Again, Eric and Eleanor were 'living out'. When Eleanor made her first visit to the butcher she was amazed when he rested his knife half way down a leg of lamb and enquired, 'Will that be enough?' Their Ration Book was not even requested and that leg of lamb saw them through three weeks.

During the evenings the flyers gathered in a local shebeen, a pub which was adopted as the unofficial Squadron H.Q. during recreation hours. It adjoined one end of the proprietors cottage while he had a grocery shop occupying premises at the opposite end. Running the farm, the shop and the bar while her brother was away in the forces was Mary, a petite brunette with dark flashing eyes. She used to entertain her bar customers with chilling realism, by telling stories about banshees and other beings of the supernatural. When presumably she had everyone sufficiently entranced to the extent that they were discouraged from venturing out into the dark of the night, she would escort Eleanor to the shop and load her up with provisions in such quantities that there was never any hint of a shortage.

When the Squadron started operating to Iceland, the crews found themselves in an ideal situation to do a little trading. Iceland provided a plentiful supply of highly prized cigarettes and ladies nylon stockings while Ireland was a source of copious quantities of fresh vegetables which were scarce in Iceland.

One evening after a party Eric was driving home from Londonderry with Squadron Leader Brolly beside him and with their wives asleep at the back. Motoring at night was not easy with the dim blackout headlights. Eric noticed a small red light waving at the side of the road. Having no more room in the car and thinking it was someone wanting a lift, he took no notice but stopped very suddenly when he heard a bullet whistle past his side window. It was the Garda checking people leaving Londonderry. When he asked what the hell they were playing at, the only reply was, 'If you hadn't stopped, the next one would have hit you!'

Nearly everybody 'lived out' at Limavady because there was very limited accommodation on the base. Personnel were given a petrol ration only sufficient to get them from their digs to the station as they had no business to be driving elsewhere. However Eric discovered that his Ford Anglia would run quite well with half paraffin and half petrol. The smell rather gave the game away but he was happy to chance that.

A few months after the Londonderry incident, Eric was taking Eleanor to a neighbouring village for a drink when, once again, he was stopped by a dim red light waving in the dark. Opening his side window, he roared at the Garda, 'If you are going to shoot me if I don't stop, you might at least have a decent light!' The policeman was very apologetic and said he was only given some red plastic to put over his small hand torch. Eric knew that attack is the best means of defence and fortunately the Garda forgot to ask him what he was doing there.

Eric's first visit to Iceland took place on 10th August 1941. The station there had run short of serviceable aircraft and had called upon Limavady for assistance. Eric set off about 5.00pm on the six and a half hour flight, landing at Kaldadarnes still in daylight. He flew in Iceland for three days and then returned to Limavady.

Flying Wellingtons across the choppy wastes of the Atlantic Ocean on anti-submarine patrols was not the 'safe' job away from the main theatre of war as may have been anticipated. Even without the provocation of enemy aircraft the operation had its hazards.

The Pegasus engines on the Wellington were not all that reliable and the aircraft would not stay up with only one engine functioning. The proliferation of aerials associated with the infant radar that was the A.S.V. created additional drag and were prone to icing up so this in no way helped such a handicap. A disabled propellor couldn't be feathered except in one circumstance. The Pegasus engines were prone to develop 'galloping oil consumption', a condition that was only revealed when an engine seized. This could cause a jerk which could make the engine shed its propellor. Such an outcome was even better than having a feathered prop, provided the departing propellor threw itself outwards, and under these circumstances the Wellington would remain airborne on the one remaining engine. If the shearing propellor spun inwards, the outcome was very different as the pilot was directly in its path and was an instant fatality. 221 Squadron flying from Limavady generally operated at 600 feet and were most of the time out of radio contact. As aircraft had to climb to a higher altitude in order to call base they frequently failed to return without explanation unless the crew managed to bail out with a dinghy and subsequently be picked up. Eric recalls only one instance during his time at Limavady when the crew of a disabled Wellington survived in this manner to tell their story.

Some pilots, like Eric, had many years of civil flying which gave them sufficient experience to learn and improvise with the equipment at their disposal. Others however had minimal experience and aircraft not infrequently went out to scan the seas for U-boats and were never seen again. Most of them were probably casualities of engine trouble, depleted fuel or lost bearings. One Irish family stopped taking service families in as boarders. After having in quick succession to console three wives whose husbands had failed to return, the emotional trauma was too much for them to consider inviting more grief into their home.

The level of inexperience among many pilots who had been propelled into the air solely because of the war was brought home to Eric when a member of another Wellington crew filled a temporary gap on Eric's aircraft. 'It's nice flying with you,' he said. 'There usually

DH86B Express Air Liner, 'The Norseman', in flight at Dyce, 1937.

A Junkers Ju-52 floatplane operated by D.N.L. Norwegian Airlines. Eric flew in these during Allied Airways' exploratory visit to Norway in 1937.

Tryggve Gran's Bleriot monoplane, 'Nordsjøen', when it was in retirement at Stavanger Airport. *(Courtesy of Jan-Petter Helgesen)*

Dr Agnes Thomson presents Eric with a sprig of lucky white heather prior to the departure of the first exploratory flight from Aberdeen to Stavanger.

Eric and Eleanor after their wedding at St Machar's Cathedral Aberdeen. To their left is Harry Reed who was to become Allied Airways' only fatality.

'Nordsjøen' attracts the gaze of onlookers on the sands at Cruden Bay prior to Tryggve Gran's historic flight of 1914.

Germany's 'Jumbo' of the 1930s, the Junkers G38, at the opening ceremony for Stavanger Airport. Note the cabins in the wings and Gandar Dower striding across the tarmac.

An Avro Anson undergoes the scrutiny of some aviation journalists.

Flightdeck of the Anson with Eric in command.

always seems to be a flap on, but everything is so calm when we fly with you!'

The difference made by Eric's experience was ably demonstrated during an air test following some engineering work on the controls of one of the Wellingtons. Immediately after taking off on the short runway at Limavady Eric discovered that the aircraft was tail heavy. The more he trimmed forward the worse it became. In the end the aircraft was so tail heavy that he had to push the stick forward with his knee as well as his hands. He was concerned that, as the trimmer had to be in neutral before putting the flaps down so as to avoid damaging control lines and as he must use the short runway, he would need his landing flaps.

Having coaxed the Wellington to a safe height he gingerly wound the trimmer back. The more he wound it back, the better the aircraft flew until, when the trimmer was in neutral, she could be flown 'hands off'. The trimming wires were crossed and were working the wrong way. Once he knew this everything was easy but a less skilled pilot would surely have been killed. When he taxied in, the rigger came up to ask how the test had gone. Eric's response was, 'Find out what is wrong with this aircraft and think up a good excuse for not being on a Court Martial.' An hour later the fitter returned very shame faced. Eric took the matter no further.

Before setting out on long surveillance flights, a visit would be paid to Air Crew Catering. Here sandwiches accompanied by giant thermos liquid containers, filled with a choice of tea or coffee, were supplied. Eric thought little of either beverage after it had been stored in these containers for an hour or more so he soon had hot Oxo included as an additional option. This deficiency in the catering now solved, 221 Squadron would frequently set off by flying northwards to the remote Hebridean outpost of St. Kilda where the local population had been evacuated in 1930. From the sheer cliffs of this small group of islands, the Wellingtons would then fan out in triangular patterns, rather like the segments of an orange, so as to scan a wide area in an orderly and systematic fashion. The islands served as a very positive identification point.

On one occasion Eric was flying in bad weather when he was asked to fly under the clouds to St. Kilda rather than just locate it by radar. A Sunderland flying boat had crashed and they were to keep a look out for any of its occupants in the water or in dinghies.

Eric asked his radar operator to keep him just clear of St. Kilda. This was done by turning the aircraft until the radar 'blip' was on one side only of the distance line. Having given Eric a suitable course he stopped looking at the ASV screen. Eric was taking no notice of his compass, his concentration focused on the sea below in the search for survivors and so neither men noticed that the aircraft had strayed slightly off course. Suddenly to his horror Eric saw the bottom fifty feet of the cliffs of St. Kilda straight ahead and approaching at speed. A violent turn accompanied by some very rude words just prevented Eric's aircraft from joining the many aircraft wrecks that lie on the islands.

Approximately 200 miles west of the Outer Hebrides, at 57° 38′ N, 13° 38′ W, lay the only other landmark to be found in this hostile stretch of ocean. The island of Rockall pierced the waters, its jagged pinnacle of 70 feet taunting the waves. The only inhabitants who could get a foothold here were the screaming gulls. The peak of this submerged mountain had a snowy cap created by the gulls droppings. It was a useful datum point for searches further out into the Atlantic but in carrying out their surveillance flights the Flight from Limavady consistently found Rockall to be at a point 10 miles away from its supposed location. The Navy soon confirmed that existing maps had the rocky pinnacle marked in the wrong place!

Because the Wing Commander, Tim Vickers, recognised Eric as an experienced pilot, he was given freedom to introduce innovations that he felt would be productive and this even extended to modifications to his aircraft.

The Wellingtons had belonged to Bomber Command and had not been designed for the air surveillance role of Coastal Command. Their fuselages had been sealed up to prevent light getting out and allowing their detection by German fighters at night but Coastal Command wanted plenty of large clear windows to enable them to scan the sea. Eric had bulkheads cut away, all the lighting on the dashboard was rearranged, and a large diamond was cut out on each side of the cabin and covered with perspex so that they could see outside.

When Coastal Command decided that it was time that a Wellington was built for their particular needs, designers from Vickers came to see Eric's modified aircraft and it became their prototype. Eric's aircraft had the word ONE underlined and below that a figure eight. The word 'the' was painted in between them in small letters so that the Wellington became known as 'One over the Eight'.

Eric's adaptions to his Wellington were put to good effect when he and his crew were detailed to stand by a lifeboat containing the crew of a sunken merchant vessel. The large windows let them keep a watchful eye on the lifeboat and its occupants while they guiltily consumed hot coffee and sandwiches. Other aircraft, learning of the incident on their radios, joined the surveillance until a French destroyer picked up the shipwrecked crew. The lifeboat was discarded as the destroyer sailed away, leaving all the aircraft to simultaneously open fire on the empty lifeboat with schoolboy prankishness. Opportunities for the gunners to engage in target practice didn't occur very often!

On 1st October 1941 Eric took his crew north to Reykjavik for a tour of duty operating from the Icelandic capital, the whole squadron having been posted there. For this posting Angus's pram had been ousted from the removal and on this occasion Eric was carrying a motorcycle on board, which he hoped to sell at a profit in Iceland — and did! Relatives left behind, feeling sorry for their men fighting an Icelandic winter as well as a war, spent their spare time in Limavady knitting balaclava helmets and thick sweaters to send to them.

On 11th October Eric's crew did get some real target practice when they caught a U-boat on the surface during an air surveillance patrol. The Wellington attacked the submarine,

Pilot Officer Edwards releasing three 450lb depth charges set at fifty feet. One exploded alongside the stern, a second went off at the conning tower and the third exploded at the prow. It was impossible to confirm a definite hit but the Navy were sufficiently satisfied to record the U-boat 'considered sunk'.

On 1st November Eric returned to Iceland after taking a Wellington to Limavady for overhaul. All major maintenance was still carried out at Limavady. Eleanor was still living there too and the Station Commander at Reykjavik, Group Captain Revel, claimed that he failed to understand why aircraft never had difficulty getting to Limavady but seemed to encounter numerous obstacles to delay their return to Iceland.

The Allied troops had arrived in Iceland uninvited and their violation of the country's neutrality caused some degree of resentment. Eric had a Sergeant Pilot in his flight, an American who had joined the Canadian forces, and he had become involved in a brawl with an Icelander. After his arrest by the Military Police he was called before Wing Commander Murdoch who heard his case and dished out a severe reprimand. Murdoch then told him, 'Go outside, put your cap on and then return on your own.'

As instructed, the Sergeant Pilot returned to the Wing Commander's office who said, 'Your case has been dealt with and it is now finished. Now tell me what really happened!' The truth then came out about an exchange of insults that resulted in blows. Murdoch replied that their job was to fight Germans and not Icelanders and it was something that he would not tolerate. Formality now removed, the American had the last word. 'I'll clock him even harder next time, Sir!' he said.

After a month a senior officer from Coastal Command visited Iceland to make an inspection. When asked why this squadron had been selected for Iceland, he said 'Well, you are so suitably equipped because of your de-icing equipment.' The adjutant quickly put him right, 'Our Wellingtons don't have any de-icing equipment at all!'

By December the Squadron were back at Bircham Newton preparing to move to a posting where they would get by without this vital equipment — the Western Desert.

ICELAND
REYKJAVIK
KALDADARNES
FAROE ISLANDS
SHETLAND
STAVANGER
ORKNEY
ROCKALL
ST KILDA
ABERDEEN
GLASGOW
LIMAVADY
BELFAST
BLACKPOOL
DUBLIN
BIRCHAM NEWTON
LONDON

# Socks, Sweaters and Sand

On 3rd December 1941 221 Squadron's tour of duty in Iceland was brought to an early close and they set off for Docking via Limavady and Silloth. The flight from Reykjavik to Limavady progressed through extremely rough winter weather and the Atlantic Ocean was festooned with fronts. Eric's aircraft was the last to leave Reykjavik and through the storm he could occasionally catch sight of a shaft of sunlight to his right. He quickly came to the conclusion that he was flying right down one of these fronts so his navigator plotted a new course about twenty miles further west. They then continued in fine weather for the remainder of the journey, while the storm sat to their left side, until they reached the Irish coast. Only then did they have to plunge back into the storm to descend on Limavady and be greeted by the other crews exchanging stories over how badly they had all been shaken about by the weather. Several aircraft had become iced up while one had even felt it necessary to send out an S.O.S. signal.

Eric can hardly have gained popularity by announcing that he had had a lovely flight through bright sunshine. However a decade of flying through often hostile skies had given him a depth of experience that the other pilots had mostly been denied.

It was not the first occasion on which Eric had astounded his fellow flyers with his weather lore. After Atlantic patrols they always undertook a met. debriefing where great interest was always expressed in the position of the fronts, wind, air temperature, etc. Eric delighted the met. men when he started bringing in sea temperatures which also had a great influence on the weather. But the source of this knowledge had them puzzled. How could Eric measure the sea temperature from his aircraft? It was easy. All he did was call up a ship on the Aldis Lamp and ask them to give him a temperature reading from the water!

The flight from Limavady to Docking was punctuated by another incident which resulted in Eric's navigator experiencing the sharp edge of his tongue. Eric was flying south from Silloth keeping, he thought, well clear of the Midlands. He was flying at about 1,000 feet below stratus cloud when a break in the cloud ahead revealed a series of barrage balloons. His aircraft was directly on course for the curtain of cables that were stretched taut and menacing below them.

Bircham Newton was a permanent R.A.F. pre-war station only two miles south of Docking. They were both fully equipped stations with hangars, stores workshops and mess buildings — valuable targets for the enemy. Since night flying attracted German intruders, it was better that night flying instruction was undertaken from an area where there was nothing worth bombing. This was therefore done from a grass satellite at Langham. To minimise attracting the attention of the enemy the use of navigation lights was dispensed with and the flare path was extinguished the moment a departing aircraft was airborne. The only additional flying for Eric that month was the occasional positioning of aircraft between these airfields. At Docking the squadron were equipped with Wellingtons adapted for their next zone of operations — the Western Desert. To Eric the most noticeable difference between these specially adapted aircraft and those used on the Atlantic patrols was the absence of cabin heating!

On 2nd January 1942 Eric set off with a crew of eight for Egypt. Most of Europe was dominated by the German forces at this time and the normal route meant skirting around areas where the potential for encountering hostile aircraft was especially high. On the first day they flew from Docking to Portreath in Cornwall. Two days later they continued via the Bay of Biscay to Gibraltar.

Gibraltar was the regular refuelling stop for Allied aircraft entering the Mediterranean. The short runway below the Rock proved tricky for some pilots. The wind had to be at a certain level before take-off was permitted and even then some pilots were in such a hurry to raise their flaps that they ended up in the sea. Eric had no such problems when he set off into the darkness on the next leg of his journey to Malta.

Tiny Malta lay at the crossroads between Europe and North Africa and the small island group had been suffering continual bombardment from German and Italian aircraft since the beginning of the War. Few convoys were able to reach the islands which were under a perpetual state of seige. Avoiding enemy aircraft during the approach to Malta called for tactics which they would be unlikely to anticipate. Cover of darkness was an essential camouflage in evading enemy fighters. Approaching from the 'wrong' direction was also helpful.

The most direct route passed right over the Italian island of Pantelleria where a German fighter squadron was based. Allied aircraft therefore followed the coast of North Africa to Cap Bon, the north-east tip of Tunisia. The German fighter squadron frequently patrolled the sea between Pantelleria and Cap Bon trying to intercept Allied aircraft bound for Malta so, as soon as Cap Bon was reached, they would turn south to make an arc around the Italian island and to evade the fighters before making their final approach to Luqa airfield, Malta.

During Eric's flight from Gibraltar to Malta he had his A.S.V. equipment in operation and therefore was fortunate to be in a position to detect a blip coming from Pantelleria as he drew within its vicinity. He descended to sea level to observe a German twin engined fighter pass above him, silhouetted against the moonlit sky. Had the positions of the two aircraft been reversed, Eric's Wellington would have been spotted and his story would probably have finished right here!

Upon arrival at Malta they taxied to what was effectively a satellite airfield to Luqa with aircraft parked haphazardly all over the place. One of the crew commented how vulnerable they all looked when, as if to emphasise his point, the air raid sirens went off prompting their speedy deplaning. As it turned out, it was one of the few false alarms to be experienced by Malta.

The next morning Eric was greeted by his friend from Brooklands Aviation, Tony Spooner. Tony encountered Eric in a poise that typifies his unflappable demeanour. Eric had been roused by another air raid warning and Tony found him standing near the airfield mess wearing his pyjamas and a borrowed tin hat, sucking his pipe and watching an aerial dog-fight in which a Ju-88 eventually went down in a plume of smoke. Tony Spooner reflected later, 'He seemed utterly unconcerned and at ease. It was a pity that he couldn't have stayed and joined us in Malta.' But that wasn't possible so Tony settled for the next best thing and that was that they should promptly renew their friendship. So he proceeded to lay on some hospitality. However they almost ended up with more than Tony had anticipated or intended. Tony Spooner takes up the story.

'I took Eric to Eddie's bar at the Monico, the favoured drinking spot of the R.A.F at Luqa. There we had quite a few draughts of what passed for beer in that desperate place. By the time that Eddie's supplies had run out, it was too late to get back to Luqa so, for the first time ever for me, I accepted Eddie's offer to take myself and Eric to Toni's infamous, but well-known, black-market hotel, the Mayfair.

'Toni was a great admirer of the R.A.F. Nothing was too good for us. At the hotel I experienced the luxury of sheets. I hadn't seen such an item for months. After a good sleep we were woken up (we were sharing a room) by a couple of 'local ladies' who wanted to know if we required their 'services'. They came with Toni's apologies that he was sorry not to have been able to offer us anything earlier as the girls had been rather busy that night! It was about 7.00am and we had mild hangovers. As tactfully as we could, we declined their kind offer. It was all very thoughtful and civilized!'

Eric's own recollection of the 'birds' encountered that night is supplemented with the meeting of another Starling. 'Starlings were rare in the R.A.F. but I remember meeting one in Eddie's Bar. He was an Accounts Officer.'

Tony Spooner commanded a unit based on Malta and he had the green light from his Air Officer Commanding to purloin any visiting aircraft to supplement their forever diminishing servicable fleet. He had 'acquired' several Wellingtons from 221 Squadron as they passed through the island one by one on their way to North Africa. Tony decided that he must let Eric keep his aircraft, although Eric recalls that a bit of swapping nonetheless took place. He arrived on the island in a nice new Wellington and departed again in a clapped-out old one needing a complete overhaul which could be provided at Cairo.

On 8th January another nine hours of night flying took the crew, now increased to eleven as additional personnel from Malta who had completed their tour of duty hitched a ride, to Mersa Matruh on the Egyptian coast. Their destination should have been Cairo but for once Eric's intuition let him down. Eric relates the incident.

'On the night flight to Cairo West, for the first time I didn't fully trust Jack Edwards, my excellent navigator, to my cost. When we hit the coast I wouldn't cut into the desert until I was sure we had hit it in the right place. Alexandria shouldn't be far down the coast to the left so I turned down the coast to check it. There was no city to be seen so I went the other way.

'I eventually came to what looked like a firework display and which later proved to be the battle of Halfaya Pass. I realised I was in a mess and broke radio silence asking for a bearing. I couldn't get a reply from anyone so for the first and only time I sent out an S.O.S. This was picked up by Malta who gave me a bearing which put me west of Mersa Matruh. At my request, they were asked to put out a flarepath for me. When I reached Mersa Matruh I saw a somewhat ragged flarepath and assumed that was the best they could do in a hurry in the dark. I made a couple of approaches but somehow it didn't look quite right. Just then the real flarepath came into view and we all thankfully landed. The ragged flarepath that I had nearly chosen for our landing turned out the be the centre of a prisoner of war camp. Mersa Matruh was a forward casualty clearing area and we spent the rest of the night on somewhat blood stained stretchers.'

Eric and his crew continued to Cairo next day and on 11th January they touched down at 'Landing Ground 39', Burg el Arab, 60 miles west of Alexandria. The total time in the air since leaving Docking had been 31 hours 50 minutes.

The posting of No. 221 (General Reconnaisance) Squadron to the Western Desert had been a matter of some urgency, or so Eric, as Squadron Leader, had been informed (and not merely to spare its flyers from the cold Icelandic skies, it having been discovered that their aircraft weren't fitted with de-icing equipment as had been supposed.) The high priority posting had been prompted by the ease with which Rommel's army was being supplied from Sicily. The objective was therefore to find a means of spotting the German convoys crossing at night so that torpedo carrying aircraft and submarines could be homed in on them. An A.S.V. equipped Wellington Squadron was the obvious answer and that was where Eric and his colleagues played their role.

Upon his arrival in Cairo, Eric reported by telephone to Air Headquarters only to receive a response of, 'It's lunch time. We're closed until four o'clock!' Having concurred with this casual rebuff, Eric reported in person at 4.30pm only to learn that no one at Air H.Q. had heard of this high priority posting of 221 Squadron. He was billetted for a couple of days, while his credentials were checked out, in an upstairs flat with some army personnel just returned from the Front. The soldiers happily kept the newly arrived airmen entertained with stories of the conditions that they had experienced in action.

On 11th January they finally arrived at Burg el Arab in time to catch the tail end of lunch. Eric was most impressed with the high standard of catering, but this was casually dismissed as only what was to be expected — since the cook had been Lady Astor's chef before the war. A neighbouring squadron also tried to put on a good show with its hospitality and a bar mounted with a brass-handled beer pump looked the height of luxury in its desert surroundings. But it was a device to kid the unknowing because, while the bar attendant was pulling the handle, a small Egyptian was filling the customer's glass from a bottle down below.

The mess at Burg el Arab was in a marquee which was unusual in that it was liberally embellished with autographs. These were souvenirs from a dancing troupe lead by Alice Delicier that had come to perform at Alexandria. A party had been laid on for them after their show in Alexandria but the troops from Burg el Arab, who had been denied a share of the entertainment, had other plans. With a fleet of trucks they abducted the whole dance troupe on their way from the theatre to the party and whisked them off into the desert to put on an unscheduled performance for the troops 'up the road'. The girls had taken it all in good humour!

Most of the stations in Egypt were hastily constructed bases in the desert which were referred to by the R.A.F. merely by a Landing Ground number since there frequently was nothing nearby from which to borrow a name. Burg el Arab was a little different in that it was overlooked by a handsome fort that had been used as the setting for the film 'Beau Geste'.

The desert around Burg el Arab was of dry clay which soon became ground into a powder and took on the appearance of brown flour with the continual passage of heavy military transport vehicles. High winds frequently blew up creating instant sandstorms with this 'brown flour'. Visibility was reduced to yards and personnel found their way around by following the telephone wires which just lay on the desert floor. A wire was followed to the 'grand junction' which was represented by the telephone exchange tent. Then, like the roads leading from a roundabout, the appropriate cable to the mess or any other destination was selected and followed to the destination.

Eric continues. 'Under normal circumstances, if we wanted to visit a neighbouring Squadron, we lined up our car using the layout of our tents or a compass. We then used a sun compass for keeping a straight line. Every vehicle had a bolt sticking up through the bonnet. A circle was painted round this bolt and all one had to do to motor in a straight line was to keep the bolt's shadow in the same place. One evening we were invited to a mess party at a squadron close by. The driver who took us there was very worried about getting back on his own. We said to him, 'You see that bright star low down on the horizon? Well, just drive towards it and it should get you back alright.' It so happened that he finished up right in the entrance of the Motor Transport Section. Later he said he never knew that astro navigation could be so accurate. We didn't tell him it was pure chance!

'People could be very stupid in sandstorms. One morning a sandstorm arose very quickly. Three airmen working on an aircraft headed for their mess not using phone wire navigation. Visibility was so poor that they went past the mess tents without seeing them. One walked and walked until, in great distress, he happened on a neighbouring outfit. Another was picked up by the army. The third walked for ten minutes then, seeing nothing but reckoning the tents couldn't be far away, did what the Arabs do — sat down with his back to the wind awaiting a clearance. Within a short space of time the wind lulled for a bit, he got a glimpse of the tents and was home and dry.'

Once Eric and his crew were settled into their station at Burg el Arab, Eric's first chore was to take his Wellington aloft to demonstrate the capabilities of its A.S.V. radar equipment, quickly convincing those in authority that it could be a vital component in aiding the war in North Africa.

Then by way of a totally different diversion, Eric went to 'Faiyum Road Kilo 17', an airfield just south of Cairo, to ferry a Miles Magister back to Burg el Arab.

By the beginning of February the squadron had settled into a regular routine of conducting aerial surveillance at night across the eastern Mediterranean, experimental flying in co-operation with torpedo aircraft, and once the art had been perfected, dropping flares over enemy shipping that had been detected by the A.S.V. so that the torpedo aircraft could move in. Ports of call along the Egyptian coast included westward flights to Sidi Barrani and to Gambut and Bu Amud, both near Tobruk in Libya. Eastward flights took Eric to Shallufa, near Suez.

However most of the flying was across the water seeking German supply ships sailing from Italy. On one occasion Eric flew a little further in pursuit of ships than he had intended. 'We reckoned we were about fifteen miles from Sicily and should start turning back. I changed over to my front aerials and to my horror the screen became a mass of blips — I was over Sicily but fortunately high enough not to have hit the mountains. The cause was one of two things. Either my Air Speed Indicator was under-reading or an unexpected tail wind had arisen. If it was the latter, I was going to have to fight my way back through a headwind and might not have enough fuel to get back. Fortunately it was the Air Speed Indicator which was at fault.'

On some flights Eric was joined by Wing Commander Alister Murdoch, an Australian who had also been the squadron's Commanding Officer in Iceland. During the squadron's Iceland posting he had instituted a weekly Squadron Parade which was most unpopular. Eric, who assesses his entry to the R.A.F. as 'via the back door', felt at a particular disadvantage when called upon to parade his Flight since he knew nothing about drilling. 'I frequently gave them the wrong order and finished up with them with their backs to me. I soon learned to roar 'As you were!' and the troops loved it.' To the chagrin of the squadron, Wing Commander Murdoch acquired a flagpole for their desert outpost and continued to have his Squadron Parades in Egypt. Then one day the flagpole disappeared and was eventually found nearly three hundred miles away at Sidi Barrani having arrived there in the bomb bay of a Wellington. The culprits were never found but Murdoch realised he was fighting a losing battle and the incident concluded the practice of Squadron Parades.

Most of the longer aerial surveillance flights were carried out at night when the German convoys, and the A.S.V. Wellingtons that were seeking them out, could enjoy some protection from the cover of darkness. Having located enemy shipping, Eric and his colleagues would pass their quarry over to the torpedo carrying aircraft to drop their devastating cargo on the vessels. The Wellingtons would provide further assistance by dropping flares over the convoys.

The potency of underwater explosive devices was well respected. Wellingtons could carry three depth charges. When they dropped them into enemy harbours such as Tobruk, the minesweepers knew that they had only to sweep up three and the harbour was safe again. Someone had the brainwave of dropping three oil drums containing nothing more lethal than sand. When the minesweepers managed to explode nothing, the Germans became very concerned and thought perhaps a new firing mechanism had been introduced. It was quite possible to design a fuse so that several ships passed over safely before the final vessel set it off. A lot of abortive sweeping was done before they dared use the harbour again, sometimes causing a delay of several days. The oil drums were much more effective than real depth charges but it was a bit hard on the crews who had to dice with death to drop barrels of sand into the sea.

Games of bluff were not uncommon. On 11th May 1942 Eric made the following entry in his pilots log, 'Sighted enemy submarine heading for Bardia. Nothing to attack it with! Damn it!' The A.S.V. Wellingtons didn't carry heavy armaments but the U-boat didn't know that when Eric nonetheless dived on it forcing it into a panicked submergence. As Eric commenced his final month of his tour of duty in the Western Desert he was able to record a personal success when he was the only aircraft aloft to locate a night convoy. He guided in the torpedo aircraft and a 10,000 ton ship went to the bottom as a result.

May was also the month during which the postman came. The Christmas mail for 221 Squadron had finally caught up with them and to great amusement its members opened parcels containing woollen socks, scarves, mittens, balaclavas and sweaters that wives, girlfriends, sisters and mothers had judged would be ideal gifts for their men in Iceland!

Eric completed his tour of duty by flying his last rostered flight on 12th June. He was ready to go home on leave when two days later he was asked to 'do one more'. A large convoy had been observed preparing for departure from Sicily and it was considered vital that not all its cargo of equipment should reach Rommel. It was also pointed out that the Wellingtons might not have enough fuel to get back, depending upon how long they had to seek and shadow the convoy, so in these circumstances they should head for Malta. 'I felt that doing another flight when the tour of duty had been fulfilled was tempting providence,' remarks Eric and indeed things were to go wrong. But not at the hands of the enemy.

From his base at Burg el Arab, Eric flew to Sidi Barrani where the aircraft was refuelled. The bomb bays were converted to carry overload fuel tanks and he taxied off with 1,020 gallons on board, of which 240 gallons were in the bomb bay. However at 60 knots, just as the aircraft was readying to leave the ground, there was a loud bang and a flash. The starboard tyre had burst and this threw the Wellington into an uncontrollable swing. Eric throttled back and braked furiously but the aircraft kept swinging until the undercarriage parted company. The aircraft came to rest in a spray of sand and to a chorus of expensive noises.

'Get out as quickly as you can', Eric shouted down the hatchway. He helped some escape through the 'sunshine roof', an opening window in the cockpit roof, while contemplating valourous thoughts of, as Captain, being last to leave the stricken ship. He was well aware that he had come to rest sliding sideways on his bomb bay which was holding 240 gallons of petrol. He admits that had he seen a flame or smelt petrol, he may have forgotten any thoughts of couragous gestures. But he had already satisfied himself that there was no leakage and the crew were all safely out when their Commanding Officer arrived on the scene.

Eric remembers him as a 'Jimmy Edwards type' as he demanded, 'Is everybody out?' The crisis moment past, Eric couldn't help playing to the gallery. 'I'll just check, Sir', he responded. This he followed a moment later with an affirmative. Rather than jump from the cockpit into the dark again and perhaps break an ankle, he put a ladder over the side, calmly collected his log books, put his pilot's cap on and reported back to the C.O. with a salute and an 'All present and correct, Sir.' Eric had put on this performance as a bit of a joke but the C.O. took it all very seriously. Later in his report he was to write 'Squadron Leader Starling, by his coolness and calmness, saved what could have been a very bad accident.' Eric's own view is that once the aircraft had started to swing he was just a passenger with no control over events from that point on. Had it been daylight he would have seen that a burst tyre was the cause of the bang on take-off and he would have kept the Wellington straight while coming to a stop, with bursts of the starboard engine. But perhaps, in his usual modest manner, he doesn't give himself the credit that is his due!

As the summer of 1942 progressed, Rommel was putting on the pressure and the Allied forces were in retreat. Eric's pranged Wellington was sitting on the ground at Sidi Barrani bristling with the antennae of its secret A.S.V. radar equipment. Concerned that it should fall into enemy hands, and with no time to dismantle it, the whole aircraft was loaded on board a truck and commenced a slow journey to safer territory further to the east. However the massive load, with the aircraft wings protruding over either side of the desert road, impeded the progress of numerous other vehicles making a hasty journey away from Rommel's advancing army. To get past the massive 'Queen Mary' truck they had to drive off the road into the desert where they would be further delayed as they got stuck in soft sand. Eric of course was far away from Egypt by this time but, with tongue in cheek, claims that it was because of his Wellington blocking the road that the retreating Allied army gave up trying to pass it and so stopped at El Alamein to meet Rommel in one of the most famous battles of the Second World War.

His 'one extra mission' having been brought to an abrupt conclusion, Eric boarded the Queen Mary at Suez for the start of his journey back home. The famous liner had arrived with troops who would shortly be thrown into action to check Rommel's advance in its culmination at El Alamein. Although her wartime capacity was 10,500 troops, she wasn't carrying anything like that number of passengers on this trip; just some 'operationally expired' flying crews, civilian evacuees from Malta, and German prisoners from the Afrika Korps who were in the firm custody of Polish troops.

Among the evacuees was a small boy accompanied by his mother. The boy was known by everyone as 'Bomber' because he had been born in a Maltese air raid shelter. The German prisoners were kept secure by their Polish guards who tolerated no misdemeanours. When the prisoners were brought up on deck, one of their number was caught spitting at a portrait of Queen Mary that was displayed along the way. A few days later the same prisoner was buried at sea having died of 'natural causes'. Who knows whether the two incidents were directly connected.

As the 81,237 ton liner set off down the Suez Canal, the ship was given several air raid warnings for the first hundred miles or so and on these occasions the passengers were sent down below. Eric and the other Coastal Command pilots spent as little time as possible below decks during these warnings. They reckoned that enemy aircraft would not attack ships with instant fused bombs because these would only blow off the superstructure. A fuse with a few seconds delay ensured that the bomb was well inside the ship before exploding, so below decks was the worst place to seek refuge.

The vessel then entered the Red Sea where they experienced an oily calm accompanied by a heavy heat haze. It was so calm that the airmen didn't realise that the Queen Mary had done a quick turn to starboard until they noticed its wake had turned 90 degrees. Immediate thoughts turned to U-boats and torpedos so they rushed to the bow of the ship to see what was happening. There they saw in the haze the outline of an island not very far away. It pleased them no end to discover that the Queen Mary, with all its sophisticated navigational equipment, had found it necessary to do a 'splitass turn' to avoid running aground.

Tracking the ship's progress proved to be a popular pastime. The air navigators had a fairly good idea of the Queen Mary's speed so they could plot the course of the ship. One navigator had astronomical tables and he would use these to get a position line by spending hours on deck trying to identify whatever star was immediately above him. As they approached Madagascar seagulls started to fly behind the ship. From their flying over the North Atlantic they knew that the maximum range of the gulls from land was about one hundred miles and this served to confirm the accuracy of their plotting. On a different voyage some Coastal Command crews had apparently been so accurate in pinpointing the ship's position at any given time, the captain thought they must have something that he didn't know about so he confiscated their maps.

At Cape Town Eric left the Queen Mary and continued home in a Dutch ship called the 'Nieuw Zeeland', eventually disembarking at Stranraer on 3rd August. While he had been in Egypt, Eric's wife, Eleanor, and his son, Angus, had been spending some of the time with a cousin at Lambourn in Berkshire. This cousin had married Arthur Saxby, a racecourse trainer, and it was here that Angus was found playing between the legs of a dangerous stallion. The horse, which struck terror into everyone else, was assuming a protective role over the infant boy. Eric collected Eleanor and Angus, had a period of leave, and then returned to now familiar territory — Limavady.

Eric took up his duties at Limavady on 22nd August 1942 as Officer in Charge of No.3 Squadron, No. 7 Operational Training Unit Coastal Command, an operational training unit for Wellingtons. No. 3 Squadron consisted of Ansons and banner-towing Lysanders. In the early months of this posting the aircraft that was to be his workhorse was therefore the faithful Avro Anson, although the opportunity for an occasional flip in a Lysander sometimes arose.

The role of this unit was to undertake the final stages of training for navigators and wireless operators. The Anson was used to take pupils on training flights across the waters of the Atlantic. The Lysanders were used for drogue-towing, the drogues being the targets in anti-aircraft gunnery exercises.

Eric believed that discipline was much more effective if it was good natured. He therefore instituted a system of fines. The fines were graded to range from minor transgressions up to acts of major stupidity. If a pilot taxied in with his flaps down (easily damaged on the Anson as they were so near the ground) a fine was paid into a fund. As the pilot deplaned, his colleagues quickly pounced on him to demand his money. Needless to say, the proceeds of this fund were used for the occasional 'squadron night on the town'.

Eric was kept busy flying pupils in the Anson so he didn't get time to take instruction on the Lysander. However one day there were no Lysander pilots available and the unit was anxious to undertake some drogue-towing. Would Eric take one up? So Eric set off in a Lysander, with the handbook clutched in one hand, to do a series of flights up and down the coastline. While gunners practised firing at his drogue, Eric was matching up control knobs with the diagrams in the handbook to find out their function. At the end of an hour's flying he was confident that he now knew the intimacies of the Lysander.

The routine of the Anson training flights brought its moments. One trainee wireless operator reported to Eric on the flight deck, 'I'm sorry, Sir, I can't make head nor tail of this message. It's all in gibberish.' The message was in code and read, 'Weather closing in. Return to base immediately!' A quick race back to Limavady just beat the encroaching storm.

It was at the beginning of 1943 that Eric moved to No. 2 Squadron at Limavady. Here he was training crews on Wellingtons, although he still occasionally flew Ansons, or Miles Martinets which were used in assimilated fighter sorties with the Wellingtons. On 25th January Eric used an Anson to fly the badly burned survivor of a Sunderland Flying Boat crash on an air ambulance flight from Limavady to Abbotsinch near Glasgow.

At another part of the Limavady airfield pilots were taught to fly, navigators to navigate and wireless operators to operate the communications. They were then sent to Eric's No. 2 Squadron to be trained to work together as a crew. In conveying the necessity of teamwork and discipline in an aircrew, Eric briefed trainees with his own assessment on the reality of how the pecking order worked in the services. 'An aircraft cannot be run by committee. If you as a crew put up a good show, it is the Captain who will get the Distinguished Flying Cross even if he was asleep in the back of the aircraft at the time. So if you put up a bad show, I'll see that the Captain is the one who gets thumped.' The point being that it is the Captain who was in charge of the crew of six, he was the boss, and it was the duty of the crew not to let him down.

Eric's predecessor at No. 2 Squadron had great difficulty in getting his pupils through their entire course in the allotted time and his reputation had taken a knock as a result. Eric encountered similar problems during his first course and he set about doing some calculations to reach a solution to the problem.

Taking into account disruptions caused by weather, Eric reckoned that if he had four Wellingtons at his disposal during the day and two at night, he could see his courses through to their conclusion. He put this assessment of the situation to Wing Commander Paddy Grennan, the Chief Technical Officer who was in charge of all the maintenance. Paddy was quite emphatic that provision of this number of aircraft would be impossible because he was short of fitters and riggers. Eric next approached Wing Commander Greenhalgh, the Chief Flying Instructor, explaining his view of aircraft requirements and their unavailability. Greenhalgh and Grennan then had a head-to-head while Eric became a bystander. Paddy Grennan of course won the argument because the aircraft simply weren't there, but it left Eric in a situation where he could only do his best but needn't have sleepless nights when courses couldn't be finished before the next intake.

In June 1943 Eric received his first command with a posting to No. 304 Ferry Training Unit on the Isle of Islay. The function of this unit was to give newly trained crews on Beaufighters additional practice on an aircraft, which was normally newly acquired from the factory at Bristol, before taking up a posting in the Middle East or the Far East, the latter identifiable by a snake painted on the side of the aircraft. Training flights would head far out over the Atlantic to give the crews experience of distance flying. Eric would kit out the Beaufighter crews, set their compasses, swing their propellors and check their guns. Eric noted his first Beaufighter flight as 'fastest yet' and this was to be his steed on Islay except for the occasional interruption of a flight in a Proctor.

During this phase one Beaufighter was lost when it crashed on St. Kilda. LX798, newly delivered from the factory at Bristol, set out from Islay at 2011 hours on 3rd June 1943 for a night flying exercise. On the flight deck were eighteen year old Sergeant Bill Duxbury and twenty year old Sergeant Stanley Thornton. The level of experience available to many of these pilots is indicated by the mere five hours previous night flying on the Beaufighter that was credited to Sergeant Duxbury. When the aircraft failed to return, an exhaustive search was mounted but to no avail. It was only after a gap of nearly three months that the mangled wreckage was spotted on Conachair, a 1,376′ crag on the northern side of Hirta. Fragments of wreckage remain there to this day as a silent tribute to the two youthful flyers.

Several other Beaufighters were written off either during take-off or landing. This was because the Beaufighter had a traditional undercarriage when a tricycle undercarriage would have been more suited. The aircraft therefore suffered from 'oversteer' and if it was allowed to swing by more than ten degrees it became impossible to hold and the pilot would experience his tail overtaking him! Aircraft would therefore frequently disappear off the side of the runway. Eric's only experience of this effect was at Aberdeen where he landed on the short runway but was able to take corrective action by turning on to the long runway at the intersection and therefore give the impression of a totally planned manoeuvre.

On one occasion Eric had to fly a Beaufighter to Benbecula to serve in a Court Martial. Rain was falling incessantly, visibility was poor and the cloud was down to 400 feet when he reached Barra Head. Eric crept up the coastline of South Uist until he reached the right headland for Benbecula, found the runway and comfortably set the Beaufighter down. Benbecula was the base for a Flying Fortress Coastal Command Squadron and they had been grounded for the last three days by the bad weather. When their Group Captain heard Eric's Beaufighter coming in, he phoned the control tower with, 'Who's that bloody maniac flying in these conditions? Send him straight to me.' 'It's Squadron Leader Starling. He's ex-Scottish Airways.' came the reply. The tower got the airline name wrong but it had the desired effect. 'Oh, that's alright then,' responded the Group Captain, 'Forget it.' Upon being told of the exchange later, Eric became aware for the first time of the high regard in which the Scottish civil pilots were held.

In October 1943 Eric moved from the islands of the Scottish Hebrides to Thorney Island, near Havant, on the coast of Sussex.

Iceland to the Western Desert January 1942

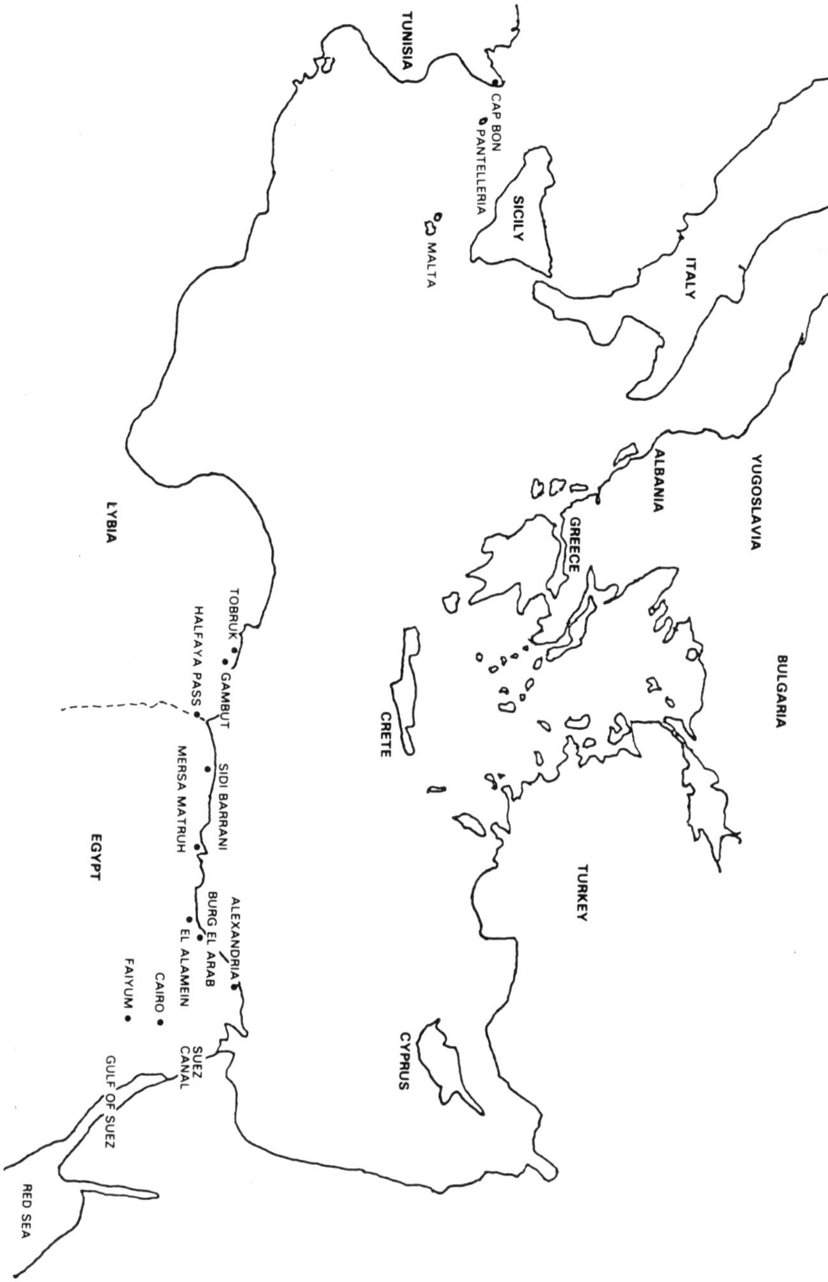


Western Desert 1942

# Bengal Beckons

Eric spent the next six months at Thorney Island. For the first month he was attached to the Air Sea Rescue Training Unit and during this time he flew the Anson and Wellington but was also introduced to a new aircraft, the Vickers-Armstrong Warwick. Eric's Warwick conversion took place at Thornaby-on-Tees, near Middlesborough

At 46,000 lbs the Warwick was the largest aircraft that Eric had flown to date but it brought with it a familiar feel since it had been a precedent to the Wellington. When it was first built it had been intended to equip the Warwick with Hercules engines but these didn't materialise on time. A smaller version, the Wellington, had therefore taken to the skies instead. Confirmation of this was to be found in the Wellington fuselage where the frames carried a series of numbers. A break in the numbers showed where the aircraft had been shortened.

However, the Warwick, fitted with two Pratt & Whitney Double Wasp engines, had now arrived. Pratt & Whitney had made a large number of Twin Wasp engines for the French but, with the collapse of France, they now found a use in the Warwick. While its outline resembled that of the Wellington, operating on Air Sea Rescue duties the profile of the Warwick was changed by the addition of a lifeboat which it carried beneath its belly. Eric's lifeboat, the first to take to the air, was named 'Eleanor' after his wife. The airborne lifeboat was being used in British waters but as it happened, no further lifeboats followed the advance party with 'Eleanor' and she was never put to use because the necessary parachutes never materialised either.

Having completed his conversion on the Warwick, Eric formed No. 292 Air Sea Rescue Squadron at Thorney Island in December 1943. Training and exercises on the Warwick followed during the next four months. Some flights in a Miles Martinet were made between Thornaby-on-Tees and Manby in Lincolnshire. But flying of a totally different and unmilitary nature came courtesy of the Group Captain at Thorney Island who owned a Tiger Moth which senior officers occasionally got an opportunity to fly.

This was Eric's kind of flying and he took the Tiger Moth on an overnight trip to Detling, near Maidstone in Kent, where he demonstrated his skill in aerobatics to his fellow aviators. A flying officer by the name of MacGrath was desperate to be taken through some aerobatics since, having completed two operational tours of duty, he felt that he had omitted an important part of his flying career by never having flown in an open aircraft. The attraction of fulfilling this ambition while the machine was being put through its paces was too good to miss.

Eric duly obliged Flying Officer MacGrath but upon becoming airborne he discovered that the aircraft didn't have an Air Speed Indicator — Eric would just have to judge his airspeed to the best of his ability. He climbed and he dived until the aircraft whistled and screamed. Loops, stall turns and spins followed until at 1,000 feet Eric planned a final loop to complete the show. However this time he went into the loop too slowly and suddenly, with his tail to the ground and nose to the sky, the aircraft came to a near standstill and with no wind his propeller stopped dead. In such a situation Eric had been told that if you put the aircraft into a dive and then pull it out suddenly, it is possible to restart the engine. Putting this theory to the test, Eric dived a couple of times. The engine had reluctantly kicked over on only one of these attempts but had refused to restart. He was now down to 500 feet and decided that it was time to execute a forced landing on the grass between the runways at the aerodrome. This was accomplished without further mishap.

Eric apologised profusely to MacGrath for the incident, explaining the technicalities of how the too-slow climbing speed had caused the petrol to sink to the back of the aircraft and so starve the engine of fuel and with no wind to keep it going. MacGrath was delighted. 'Ah, at last I can say 'There I was, upside down, no speed on the clock.', he beamed. Eric took the Tiger Moth straight back up again and this time completed a perfect loop.

At Thorney Island Eric and his colleagues had a ringside view of the German blitzing of Portsmouth. One night after a film show at the camp cinema, Eric took a short cut right across the airfield to the officers mess. When half way across, a German flew overhead on his way to Portsmouth. The anti-aircraft guns started firing at it and shrapnel starting falling around Eric. He had no tin hat and the nearest shelter was a long distance away. 'I have never felt so vulnerable in my life,' he recalls. It was obviously time to get away from Thorney Island.

On 4th April 1944 Eric set off from Portreath in Cornwall to take his squadron on active duty in India. Sending him off was Eleanor who was expecting their second child. Eric arranged for her to stay with his sister at Castle Douglas and it was here that Zoe was born on 10th May. Had the baby been a boy he was to have been named Robert after Eric's father. Deciding on a girl's name wasn't so simple and required a visit to the Douglas Arms to give the matter some serious discussion. Eric mentioned his affection for a cousin called Zoe which Eleanor noted was Greek for 'life' — so the decision was made. When Eric was to see Zoe for the first time she was nearly two years old.

Eric's route to India took him to Rabat Sale in Morocco, Castel Benito near Tripoli, Libya, and then to Cairo West. The journey was all undertaken during daylight indicating the change in conditions in the Mediterranean and North Africa that now prevailed. The crew had a four day stop at Cairo while Eric indulged himself with a visit to the Pyramids, an experience denied him during his tour of duty in Egypt. From Cairo he flew to Habbaniyah in Iraq where they took off only to make a quick return — they had left with the petrol cap off!

The next stages took them from Habbaniyah to Karachi via Sharjah in Saudi Arabia. At Sharjah another slight delay occurred to permit repairs to the engine cowling of the Warwick while Eric was entertained by the British Consul who resided in a fort that resembled a French Foreign Legion garrison. Between Sharjah and Karachi Eric climbed to 10,000 feet to escape the heat. 'In Coastal Command our average flying height was 600 feet. It used to be said that we got dizzy above 1,000 feet. Mentioning the height at which we had flown to my crew, my tail gunner exclaimed, 'So that explains it. Getting in and out of the rear turret I started getting short of breath and I thought I was ill.' He was greatly relieved when he knew it was height that was causing it.

The journey to India had taken 37 hours 40 minutes of flying time over a route of 5,523 nautical miles at an average speed of 146 knots. Arrival in Karachi signalled the start of an Indian sojourn that was to be a lasting experience for Eric, particularly as he comments that it gave him the chance of becoming close friends with the Indians. He contrasts it with his posting to Egypt where the opportunity to visit an Egyptian in his home never arose and the 'visitors' never really gained an opportunity to get to know their hosts. The term 'India' is used loosely since many places to be mentioned are in what are now the independent nations of Pakistan, Bangladesh and Burma (Myanmar).

From Karachi Eric flew to Air Headquarters at New Delhi for two weeks of briefing then, at the end of April, he continued eastwards via Allahabad to Jessore from where he would operate. Vital statistics for Jessore noted that in the first six months of 1944 deaths outnumbered births by 28,517 to 16,606. Of the deaths 20,380 were attributed to malaria. But Eric found it to be a hospitable place, although some servicemen took unfair advantage of this hospitality resulting in embarrassing consequences. He tells of the incident when describing his introduction to Jessore.

'I arrived at Jessore as a very new Wing Commander running No. 292 Air Sea Rescue Squadron. Within a few weeks Air Commodore Mellersh had made me Station Commander of Jessore as well. Jessore was an all Indian town which at one time had a large British population attached to the jute industry. They had all died out through some plague but had left some fine buildings and a good tradition. In the centre of the town was a beautiful church surrounded by lawns and gravestones. It looked exactly as if it had been lifted from an English village. The local Indians, although Hindus, had carefully looked after the church. The R.A.F. took it over from them for the duration. This town had an Indian District Commissioner and a large University. The students naturally resented our presence.

'I made great friends with one of the local lawyers who spoke perfect English although he had never been out of India. Our friendship started when I was exploring the town with a colleague of mine, Arthur Hutchings. We were looking through the garden gate of a very nice house with a private Hindu temple at the bottom of the garden when the owner, Moni Mozumder, said, 'Come on in'. This started a lasting friendship. We used to go round of an evening drinking coffee on his veranda. He had a little six year old daughter with great big dark eyes. She used to appear from somewhere and walk round us at a safe distance inspecting us with these eyes opened wide. After she had done this a few times, I put my arm behind me and made a grab for her legs. She ran off with a squeal but the next time she came a bit closer. Before long she was jumping on my knee the moment I arrived and was going through my pockets to find what I had brought for her. When I was in Egypt I always had a feeling of frustration. I had so many interesting people around me but I never saw them in their homes with their families nor had the opportunity to explore their thoughts.

'In India, through Moni, I got to know many locals. I quickly discovered that the ones I liked best were the ones who said openly, 'Quit India! Now come in and meet the family.' The ones who swarmed all over one were the ones waiting to stick a knife in one's back. The local G.P. was a very likeable chap who had done a stretch in prison for civil disobedience. His wife was an efficient bustling little woman running all the local committees. So like many women at home. The camp had an official photographer whom I chucked out because he had been fiddling the books. There was an Indian photographer in the town who used to lend his dark-room to some of us. As Station Commander, I couldn't approach him directly, but I asked Moni to let him know that if he applied for the job he would get it. He was a proud man ashamed of his compatriots who ran round the camp like a lot of vultures. 'Let the R.A.F. approach me and I will think about it', was his reaction.

'One night a party of drunk members of my Squadron were attacked in the brothel area by students and one of my officers got a brick in his eye. Having deposited him at sick quarters, a gang of Sergeant Pilots decided to go back and even the score. They broke into a house by the brothel area, smashed up a private temple on the ground floor and went up through the house creating havoc. Unfortunately for them, they had picked the wrong house. It was the house of the Retired Chief of Justice of Bengal, a very distinguished Indian who was away at the time. The only people in the house were his wife and two young daughters who locked themselves in an upstairs room. I admired this Indian woman because when they broke into the room in which she was hiding, she kept her cool, explained who she was and suggested they left as soon as they could. Fortunately, they did.

'I consulted Moni the next day and he suggested I should go and see the Mayor. This I did with the officer with the black eye. I walked in to find the Mayor surrounded by all his councillors, the Retired Chief Justice, his wife and children. The Mayor delivered me a lecture to the effect that if this was British behaviour they would prefer to host the Japanese. I knocked his feet from right under him by being very apologetic about the whole incident and promised to take the appropriate action. They were used to the British Indian Army who would probably have insulted them back and told them to get stuffed.

'The whole atmosphere changed. My colleague found himself seated with the five year old daughter on his knee. The Judge said he looked on me as his wife's brother — an oriental way of saying that he would trust me with his wife — and I had an invitation to dinner with him the following week.'

The fuel dump at Jessore was a square, walled by fifty-gallon barrels of petrol. Inside were stacks of petrol in four-gallon drums. One evening as Eric was having his 'canvas bath', an Indian Army Officer came in and in a very excited voice said that the locals were holding a religious procession and it was their intention, no doubt coloured by the growing national awakening, to throw their torches into the petrol dump as they passed by. In order to prevent it, the Army asked Eric to issue them with rifles. 'Not blooming likely,' was Eric's response although phrased more directly and less tactfully. Instead he got the Fire Section to bring their fire engine into the middle of the petrol dump, instructed them to squirt water in all directions and if any Indians came near, to see that they got their full share. A good time was had by all and the torch procession kept well away.

At Jessore Eric took the opportunity to observe many aspects of human character. The chaplains in the Forces didn't escape his scrutiny.

'Until I went overseas, I used to accept padres as part of the furniture. It was then that I began to wonder about them, especially when I found that the troops confided in the doctor rather than the padre. Doc Tait of 221 Squadron had his ears very much to the ground and, for example, would come to me suggesting that a certain crew weren't working well together and a change might be advisable. Doctors are interested in one's failings and provide help rather than preach. So many of the padres were snobs enjoying their officer status to the full. The 'erks couldn't take their patronising attitude. I felt strongly that padres should be without rank so that they could rotate round Officer's, Sergeant's and Airman's Messes.

'I then began to wonder what they were doing in the Services in any case. The Bible says, 'Thou shall not kill' and 'Thou shalt love thy neighbour as thyself'. To my logical mind, a man of God wearing the uniform of an organisation busily killing its neighbours didn't make sense. I found so many of them a muddled thinking lot. The padre at Thornaby used to sit outside his beautiful little chapel with an airgun shooting the birds. The padre at Jessore used to amuse himself by seeing how many wild dogs he could run down on the two hour drive between Jessore and Calcutta. I was surprised to hear a member of the Royal family recently ask the Almighty to bless a nuclear submarine and all who sail in her. I was even more surprised to find that an agnostic like me appeared to be the only person to consider the request an obscenity.

'In India, the Christians I really admired were the Salvation Army. They were so 'down to earth'. They looked after one's physical needs before preaching. Rank or status meant nothing to them. Shortly after I arrived there, I was driving into Calcutta from Jessore with Bob Farman, our Engineer Officer, and half way there we reached a scruffy Indian village with a pontoon bridge over a river. The bridge was opened up about eleven at night and if you were late back you could be trapped on the wrong side of the river. Bob Farman said 'Let's stop for a cup of tea.' 'What, here?' I asked in surprise. 'Yes, follow me.' I then saw behind some trees a neat trim European type bungalow. We were met at the door by a very nice grey haired lady who took us into her parlour and served us with two cups of tea in attractive china cups. When I asked her how much she charged, she left the amount to me. As you have already guessed, she was Salvation Army. She was one of the refugees who had walked out of Rangoon and she had been posted to this village. She said she was the only foreigner in the village. She had turned her church hall into a dormitory where service personnel, caught on the wrong side of the river, could spend the night.

'I called in regularly and, finding that she missed fancy cakes, I always brought some back from Calcutta. When I asked her whether she could do with any assistance, she said she needed more beds and that tea was in very short supply. This I remedied because, after all, she was doing a very good service for us. Shortly before we left Jessore, a driver came into the mess one evening saying that he had called in at the Salvation Army on his way back to find the good lady was desperately ill with dysentery. I immediately sent the station ambulance to pick her up and take her to Calcutta. The Salvation Army Headquarters in Calcutta had no idea that she was ill.'

For the full gamut of local colour, it was necessary to look no further than the local market. One stall at Jessore which attracted Eric's attention was run by a young intelligent looking man who painted Indian gods in bright poster colours and sold beautiful examples of his work. He spoke good English and always worked sitting cross-legged on his stall. Eric learned from a friend of the artist that he had previously been a schoolmaster. One day on his way home from school he had been attacked by the devil and from that time he had been unable to walk. Eric's own guess was that he was a victim of polio. He got the artist to paint a series of Hindu gods on one side of air letter cards. Eric wrote a little story on the other side about the god in question and then sent them to his son, Angus. He was surprised to discover over forty years later that Angus had preserved these cards as special mementoes.

Sitting cross-legged among the vegetables that he was selling from a stall in Calcutta was a man, Indian in dress, behaviour and language. Yet physically he was a thickset, fair man with a large walrus moustache and looked just like the 'Old Bill' of the 1914/18 war. In Jessore, one of the small Indian boys seen playing around the place was fair with blond hair. Both were little gifts that the British Indian Army had left behind.

The Warwicks quickly became a disappointment at Jessore. The Bengali atmosphere was hot and humid and this soon caused the fabric of the wings to rot. Eventually they were replaced by Consolidated Liberators but until that time the squadron was getting very little flying to do. It was as a consequence of this situation that Eric was summoned by Air Commodore Mellersh with the news, 'We want Group Captain Sisson, who is running Jessore, to open up another airfield. As you're not terribly busy, you can take over running Jessore as well as your squadron.' And so Eric became Jessore's Station Commander.

This additional authority stood Eric in good stead because also stationed at Jessore was a very arrogant 'cloak and dagger' Squadron responsible only to Churchill. 'If their Commanding Officer, Wing Commander Edwards V.C. had been made Station Commander, I know he would have made us swap messes as we had the best one. As their main job was dropping agents into enemy territory, they also had a parachute training school. Their instructor, Major Thornton, delighted in waiting until some officer had rather too much to drink when he would say, 'How about a jump in the morning?' If the officer foolishly agreed, he was trapped because a jeep would be there to collect him for the jump first thing next morning.' He didn't catch Eric out like that as he had been forewarned.

No. 292 Air Sea Rescue Squadron was also a very scattered operation. Eric's base was at Jessore. New aircraft were arriving for him at Karachi. He had Sea Otter amphibious aircraft operating along the Arakan coast of Burma with the Fighter Squadrons. Additional Sea Otters were operating in Ceylon. The result was that a tour of his Squadron meant that Eric had to undertake a 6,000 mile journey. Air Commodore Mellersh, Eric's Air Officer Commanding, had known him as a civilian pilot before the war and proved to be a good ally. Eric managed to talk him into giving him a war weary Hurricane to get round his scattered Squadron. He omitted to tell the Air Commodore that he had never flown a Hurricane. Perhaps little regard was paid to such details — Eric frequently had to complete forms for H.Q. which he suspected were rarely read so that on one occasion, against the question 'On which aircraft are you most experienced?', he entered 'Avro 504K' — the prototype of the 504K was built in 1913!

'In due course a Polish ferry pilot arrived with my Hurricane. It had come from an Indian Squadron who used Hurricanes to strafe the Japanese with their eight guns, four in each wing. I had no trouble flying it. The controls were light, very similar to a Tiger Moth, and it used to glide in to land in a very docile manner. There were two things I didn't like about it. It was completely unstable — fine for a fighter pilot but irritating when used as a taxi. If you looked down to check your position on a map, on looking up again your nose was either pointing at the sky or the ground. It was like flying a seesaw. The other thing was the smelly heat which came off the engine. That part of the world was hot and smelly enough without adding to it. In revenge, I had painted on one side of the engine cowling, 'The Bengal Bitch'. On the other side I had painted, 'Hot and Smelly'. I had all the guns removed which left sufficient room in the wings to carry crates of beer one side and my camp kit on the other.

'I had been warned that it had a bad high speed stall. Many pilots were killed by it flicking over on its back on a gliding turn when coming in to land. I went up to a safe height and, with wheels and flaps down, started to do gliding turns. If I got a bit slow it just flicked over and continued turning the other way. I eventually did a tight gliding turn rather slow and the aeroplane did a flick roll. The danger about a Hurricane was that it felt so docile but would turn on you if you didn't watch your speed. When I landed, someone remarked on the flick roll I had done. 'That wasn't a flick roll, it was a gliding turn!' was my reply. The Harvard, which later replaced my Hurricane, had similar habits. When looping a Harvard one had to be careful not to hurry it round or it just rolled off the top and one could do nothing about it. To the uninitiated it meant one never completed the loop because at the top when one was upside down the aeroplane didn't like it and rolled itself the right way up.

'I would add here that I cruised the Hurricane at 190 m.p.h. and had external long range tanks fitted. An unexpected problem I had was that, after I had done only twenty-five hours flying in it, I found the tyres had worn right through to the canvas. We put a new set on and it did the same. The tracking was checked but everything seemed alright. Shortly afterwards I bumped into a Hurricane pilot and told him of my problem. 'Are you taking off at full throttle?' 'Of course,' I replied. Having spent my life flying heavy multi-engined aircraft my throttles were always pushed right to the limits on take-off. He explained that the Hurricane was a small plane with a large engine and a large propellor. Not having a second engine to counteract torque, there was a battle going on as to whether the aircraft would stay level and the propellor go round or vice-versa. One had to use a lot of rudder to keep straight and the aircraft 'crabbed' down the runway scrubbing the rubber off the tyres. 'Take-off with about two-thirds throttle, open up when airborne and you will be alright' — and I was.

'My last trip in the Hurricane was right across the widest part of India from Jessore in the east to Karachi in the west via Allahabad and Jodhpur and totalling eight hours and twenty minutes flying. From Jodhpur to Karachi was across the Sind Desert. I never trusted single-engined aircraft so I didn't go straight across but followed the railway so that I would be near rescue if the engine failed. I arrived safely but the story was different when I got ready to return. The Merlin engine is liquid cooled. Water and glycol prevent freezing. The moment I started it glycol squirted out. The leak was repaired but it squirted out in another place. After the third time it happened the Chief Technical Officer said he wasn't letting me fly it across the Sind Desert again and grounded my Hurricane. They wouldn't repair these old type aircraft and had instructions to scrap them when grounded. So that was that. The sequel to the story is that one of my squadron met some fitters from the Indian squadron. It seems that someone had put acid in the radiator by mistake! The cooling system was quickly washed out and the aircraft disposed of, unfortunately for me. How lucky I was that the trouble came when I was on the ground.'

The demise of the Hurricane had taken place while Eric had flown down to Colombo at the controls of a Beechcraft Expeditor. He had returned to Karachi as a passenger in an Avro York and just to top off the variety, he now found himself having to get back to Jessore in a Sea Otter — Karachi was where these delightful amphibians were assembled. Just when Eric was getting used to the idea of being without his Hurricane taxi, he was unexpectedly presented with a brand new Harvard, an excellent aircraft to fly and with the bonus that it could carry passengers.

When the Squadron had arrived at Jessore, Bob Farman, Eric's Engineer Officer, had reported to him that in the Station Transport Section he had found two bright young Indian lads. One was a fully qualified aircraft fitter and the other a fully qualified aircraft rigger. A fitter looks after engines and the rigger tends airframes. Bob thought that they were being grossly ill-used maintaining motor transport through prejudice on the part of the previous Squadron. Bob Farman was so impressed that, when Eric had his Harvard, he made them responsible for its care. Eric soon found that when he taxied in, even if he had been away for some time and arrived back unexpectedly, they were always there to greet him.

With the arrival of the Liberators to replace the disintegrating Warwicks, No.292 Squadron was engaged with these in accompanying Bomber Squadrons on sorties to Singapore. The Bombers would frequently get shot up and have to limp back with leaking fuel tanks, etc. Eric's Air Sea Rescue Squadron would escort them back and were therefore on the spot should a disabled bomber have to ditch into the sea. One aircraft that went into the sea in the Bay of Bengal, 1,000 miles from land, was the beneficiary of a life-raft and survival equipment within ten minutes of the incident as a result of this system. The Liberator then 'homed in' a submarine which picked up the crew of the ditched bomber. The Sea Otter single engine amphibious bi-planes performed a similar supporting role to Fighter Squadrons operating along the coast of Burma. The Liberators also ventured across these waters and on 15th December 1944 Eric records a flight lasting ten hours and fifteen minutes in which he was searching for the crew from a B29 Bomber off Diamond Island near the mouths of the Irrawaddy River, 150 miles from Rangoon.

Eric had been a keen photographer from an early age and had been secretary of the photographic society at Bishop's Stortford College. But his hobby was interrupted off the Burma coast. While at Aberdeen Eric had saved avidly to purchase a Leica camera for the considerable sum of £25. He then recouped some of this investment by always having it on hand when he flew and he conducted a regular business selling photographs of the frequent shipwrecks that occurred, these being not uncommon particularly in the Pentland Firth. He was always first to have his photos back at Aberdeen where the Press & Journal were glad to benefit from this service. However the valued Leica was to disappear while on loan to Pete Almack who was in charge of the flight of Sea Otters in Eric's Squadron. Pete didn't abscond with this valued piece of equipment — he was shot down by the Japanese and when he was eventually rescued it transpired that the Leica had been shot off his neck from which it had been hanging.

As the pressure mounted against the Japanese occupation of South-East Asia, Eric was charged with opening a new airfield at Agartala, becoming its Station Commander. Agartala is in the Indian state of Tripura and on the frontier of the modern nation state of Bangladesh. But in 1945 it was Agartala's proximity to the Burmese border that was of strategic significance.

Eric had taken to Jessore which he described as 'a nice town'. When he was asked to take a look at opening the field at Agartala, which had previously been used by the Americans, it was described to him by contrast as being 'pretty basic'. When Eric got to Agartala he discovered that the Americans had shipped every last piece of furniture out in a Dakota and the place had been left bare. He reported this back to Mellersh, Eric's commanding officer, who expressed his sympathies with a 'You'll have to do the best you can, Old Man'.

Eric paraded his squadron at Jessore to advise them of the situation at Agartala. 'It'll be alright for us officers,' he joked, 'since we have our camp kits. But if you feel that you can take anything with you that will be useful, I'll definitely be looking the other way.' The squadron seemed to look on the whole thing as an adventure as they set out for the rail journey to Agartala. Beds were being loaded on to the train to be promptly removed by the Military Police, but somehow they found their way back on to the train again and they 'escaped' with a surprising amount of extra luggage.

Eric had sent Squadron Leader Pollock, his second-in-command, up to Agartala in advance of the main party and he had scoured the area for additional equipment to help make the place appear bearable. Eric was at Agartala to see how Pollock was getting on when a visit was paid by Air Commodore Mellersh in his private Hurricane. Eric invited Mellersh to join them in the small All Ranks mess with the adjoiner, 'I hope you've brought your knife, fork and spoon with you?' A knife, fork and spoon was borrowed for the Air Commodore but he had got the message. That night he flew to Calcutta where he called together the Stores officers with instructions that they'd better see what they could do 'to help Eric out'.

Shortly after Eric's move to Agartala, Mellersh was replaced by a South African Brigadier. Eric was quickly summoned to see him and thought that he must be extremely keen to wish to meet his squadron and station commanders so soon. But it turned out to be the brawl at Jessore of six months previously that had yet to be put to rest. Eric was ushered into the Brigadier's presence. 'Glad to meet you,' said the Brigadier. 'I have just had a report from Delhi about the incident you had at Jessore. They say that as you were Station Commander you must accept some of the blame and I must reprove you. I think it is a load of rubbish but you are reproved anyway!' An R.A.F. reproof is an expression of disapproval but it isn't entered on the recipient's record.

This was a stance which Eric too was soon to adopt when he heard through the grapevine that one of his Flight Sergeant Fitters, whom he held in the highest regard, had lost his pay book. This was a chargeable offence in the R.A.F. since the lost pay book could conceivably fall into the wrong hands. Eric let the matter ride for awhile but eventually had to phone the Maintenance Section to see if the book had been recovered. 'No,' was the answer, 'the Flight Sergeant is very concerned about it. Do you want to speak to him about it?' So the Flight Sergeant Fitter was put on the line and Eric went into great detail in explaining that he would have to be brought to him on a charge and that he would have to reprove him as an expression of his displeasure. 'Do you understand?' asked Eric. The Flight Sergeant Fitter answered in the affirmative so Eric said, 'Well, that's you reproved. I'll make out the paperwork.' 'He would have been my willing slave after that if I had wished it,' recalls Eric.

Eric didn't lose contact with friends that he had made in Jessore and frequently he would visit Moni and his family for a week-end. 'The household consisted of Moni, his wife Luxmi, his ten year old son Gora and little Gopa, my favourite. In addition there was Shiuli, a niece, and various relatives. There seemed to be a bed in every room. As there was no social security or dole in Bengal, they used to run joint families so that together they could support any member being out of work. Moni was 'head of the house' and ruled the roost. He said one lost a certain amount of individuality by this system but it was worth it for the benefits. On the veranda outside my bedroom they made a temporary bathroom and when I was dressed the whole family came into the room where we had breakfast. I was once invited to an evening meal with my Indian Squadron Leader Admin., an Indian by the name of Ghoshe. It was a house where the wife was in purdah (seclusion). At the foot of the stairs to their flat, Ghoshe called out, 'Is the coast clear?'. This, he explained, was to let the wife get out of the way. At the meal, it was explained to me that the correct thing to do was to praise the food in a loud voice so that it could be heard by the wife in the kitchen. This I did and it was translated in a loud voice and we could hear laughter coming from the kitchen. Ghoshe was an interesting type. He was a Christian Bengali with a Punjabi wife so they conversed in English since they couldn't speak one another's languages.'

The Indians were a gentle people but Eric witnessed that they could display another side to their characters when the need demanded it. 'We used to have people called Enrolled Followers. They were peasant Indians who did the dirty work for us. After a while they would get fed up with the job and go back to their villages. They were hauled in by the Military Police and brought up in front of me as 'Absent Without Leave'. It was a most unsatisfactory process as I couldn't understand what they were saying and they couldn't understand me. When Ghoshe joined my staff he used to say to me, 'If you leave the office for a few minutes, I think I can sort it out. Will you please hold my wedding ring?' When I returned he reported that he had sorted it out so I said, 'Case dismissed.' He gave them a good walloping, but he wouldn't do it wearing a ring, and they understood rough justice. The wife of the District Commissioner had been insulted by one of our bearers so he was brought in by the police and spread-eagled on the table. The District Commissioner personally thrashed him with his belt and told him to leave the town and never return.'

As the Japanese were pulling back from Burma and invasion of Rangoon became imminent, Eric found delivering mail to the Navy a suitable excuse to fly there in a Sea Otter from Ramree. When they reached Rangoon they flew over the prisoner of war camp in which Allied troops were interned to see the message 'Japs Gone. Remove Digit!' painted on the roof.

On 4th May 1945 Eric therefore landed the Sea Otter in the harbour and taxied in to become what was believed to be the first Allied aircraft into Rangoon. Eric takes up the story.

'Having moored up at the harbourside, we commandeered a hut close by to use as an office. I was flying with Squadron Leader Almack and his crew. He was in charge of my Sea Otter Flight. He saw some locals looting furniture from a house and roared 'Bring it here!' at them. They meekly came along and that is how the office was furnished making us ready for business.

'Rangoon is the only town where I have seen the streets strewn with money. The Japanese having gone, people were looting the banks and discarding bundles of now useless Japanese occupation paper money. A very charming Chinese family offered us hospitality. They gave us a very good meal with a lot of drinking of toasts and I remember sleeping the night in their shop. I had with me my travelling campkit consisting of two essentials, a campbed and a mosquito net. The campbed made one independent of the type of ground. I have slept comfortably on my campbed on wet muddy ground under the wing of an Auster from which hung the mosquito net.

'Having seen Almack and his crew settled in and ready for Air Sea Rescue duties, I got them to drop me at the airport at Mingaladon. I hadn't been there long before a Wing Commander turned up in a car. He had the job of opening up airfields to coincide with the advance. He gladly handed the airport over to me and went on his way. I spent my time shredding parachutes to mark bomb craters on the runway. Some South African officers turned up and that evening I helped them drink some Indian brandy they had. The rumour was going round that Germany had thrown in the sponge so this was a wonderful excuse for a party. Some time during the next day an Expeditor landed with some of the Press. I persuaded the pilot to drop me at Ramree on his way back and from there I flew the Harvard home to Agartala.'

In July 1945 it was decided to close down No. 292 Squadron which Eric duly did having been its only Commanding Officer from the time of its formation. He was then attached to R.A.F. Headquarters Burma, which was located in Calcutta, as Air Rescue Officer. While there he discovered that there was a Sea Otter waiting to be ferried to Ceylon so he elected to fly it. Eric went to Jessore, where the aircraft had to be collected, to find that the crew hadn't turned up. He was quite happy to fly alone since service protocol required sergeants to use the sergeant's mess and officers to use the officer's mess and this resulted in much wasted time at each stopover point while crews were rounded up from different parts of an airfield. The trip was spread over three days and routed via Cuttack, Vishakhapatnam, Gannavaran, Madras and Kankesanturai to Ratmalana Airport at Colombo.

At Colombo Eric decided to visit Air H.Q. (Personnel) since he had never received a proper posting to R.A.F. H.Q. Burma. 'What on earth are you doing here?' was their response, 'You should be on your way home. Your release number for demobilisation has come up.' Eric didn't need a second telling. After one final fling, a thirty minute flight in his 40th aircraft type, a Supermarine Walrus amphibian, he rapidly lost interest in the war, returned to Calcutta for his kit and then crossed to Bombay to board the Empress of Bermuda to sail home.

India was suddenly in the past, but a small reminder of the service life there was to linger for some time. Jessore was the most malarial base in the R.A.F. and the sergeant's lines were referred to as Death Gulch. This was because it had an Indian village, plagued with malaria, on one side, a stagnant river where the mosquitos bred in the middle and a feast of sergeants blood on the opposite side. D.D.T. had recently been developed and I.C.I. experts arrived to experiment on the Jessore base. Buildings were sprayed inside and out, from the ground and from the air and this did reduce the malaria. The servicemen were given Mepacrine to take daily. This was an analine dye and turned everyone yellow all over except for the eyes. Eric had forewarned Eleanor of his appearance before he reached home but she was still shattered to see a tall, scraggy, yellow husband get off the bus. Eric had lost weight during his two years in India and he was made fully aware of this during the voyage home. Before he had left for the East, Eleanor had presented Eric with a signet ring to wear as a wedding ring. It had survived numerous tours of the Indian sub-continent but during the homeward sea journey it decided that it had had enough. Eric was playing with the medicine ball, a large heavily stuffed leather ball. He threw the ball at someone but as the ball slipped from his grasp it took the ring with it. He was horrified as he saw the treasured ring roll across the deck towards the side of the ship, but fortunately it ended up in the scuttles from where he was able to retrieve it.

While his weight had fallen in India Eric had escaped the scourges of malaria, dysentery and hepatitis. This he puts down to being born a Capricorn under the sign of the goat — in India he would see sickly dogs and cattle foraging through the filth but in their midst would be goats with beautiful shiny coats looking the picture of health. The effects of the Mepacrine however took months to wear off.

Eleanor had taken a furnished house in Braemar called 'Alt na Clunie' and it was from here that Eric was to plan a resumption of a peace time career. Pursuing his bus to Braemar was a letter addressed to 'Wing Commander Starling, Alt na Clunie, Scotland'. It arrived without delay with 'Try Braemar' written across it. Eric was most impressed that the Post Office had realised that the Clunie flowed through Braemar and it was worth a try. Perhaps the letter symbolised a return to a part of Scotland where there is no anonimity, especially for a pilot who had been responsible for bringing many of its communities even closer together through the development of air travel. Was there further work to be completed in this respect?

If any prospective employer needed convincing, Air Vice-Marshal Mellersh summed up Eric well when he wrote:

'Wing Commander E. A. Starling served under my command in Bengal for some months during the war against the Japanese. He commanded a Squadron and later a Station in my Group, and I therefore had ample opportunity of judging his ability.

'In addition to his outstanding capability as a Pilot, Wing Commander Starling ran a first class Squadron with considerable ability. Since he was not a 'Regular' Officer of the Royal Air Force his methods were, on occasion, somewhat unorthodox, but the results he achieved were excellent. I regarded him as a very capable and enthusiastic Squadron Commander.

'In due course Wing Commander Starling's Squadron was moved further forward to an isolated airfield. He took over this airfield as Station Commander and, despite his limited resources, made a very good show of it from the administrative point of view. In that part of the world we were called upon at that time to make 'bricks without straw', and the success achieved by Wing Commander Starling stamps him as a man possessing considerable administrative ability. He succeeded in building up an efficient and happy Unit and to do so required the exercise of initiative and the powers of improvisation of a high order.'

Flight Officer Jack Edwards and Eric reflect beside their Wellington, 'One over the Eight', in Iceland after sinking a U-boat in the Atlantic.

A Wellington, with the early ASV radar system, in flight. Each aerial was an ice-trap and the ASV Wellingtons became nicknamed 'Sticklebacks'.
*(Crown Copyright, Imperial War Museum)*

Far out in the Atlantic, but not quite where it should have been according to contemporary maps, the 70 feet high pinnacle of Rockall.
*(Crown copyright 1992/MOD reproduced with the permission of the Controller of HMSO)*

Arundel Castle in war time camouflage as part of a North Atlantic convoy.

The bomb ravaged streets of war torn Malta. During his visit here Eric lost his nice new Wellington to have it replaced by a 'clapped out' old one.
*(Crown Copyright, Imperial War Museum)*

Eric's Wellington homes in on a U-boat in the North Atlantic. The Navy recorded it as 'considered sunk' after this attack.

Bird's eye view of the runway at Kaldadarnes, Iceland with the camp beyond.

Eric sucks on his pipe during a moment of relaxation on some sandbags at Sidi Barrani.

Airborne lifeboat 'Eleanor' is fitted to the belly of the Warwick at Thornaby prior to departure for India.

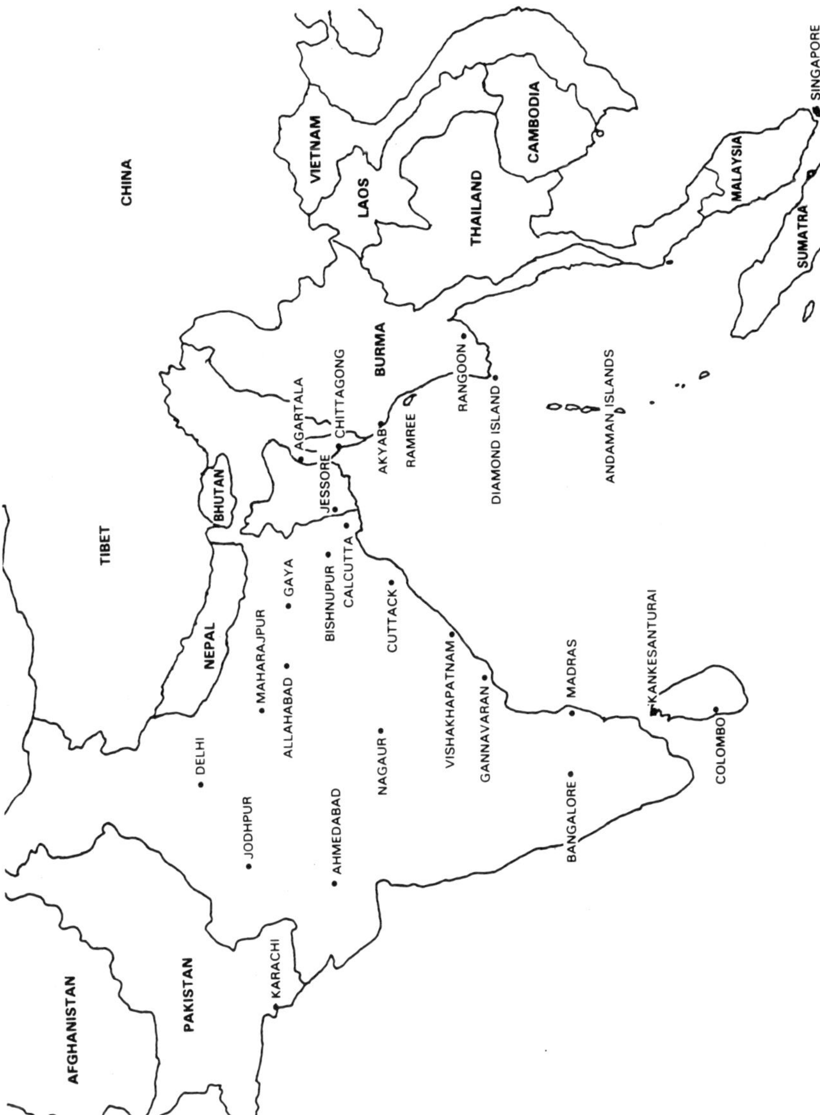


India 1943—54

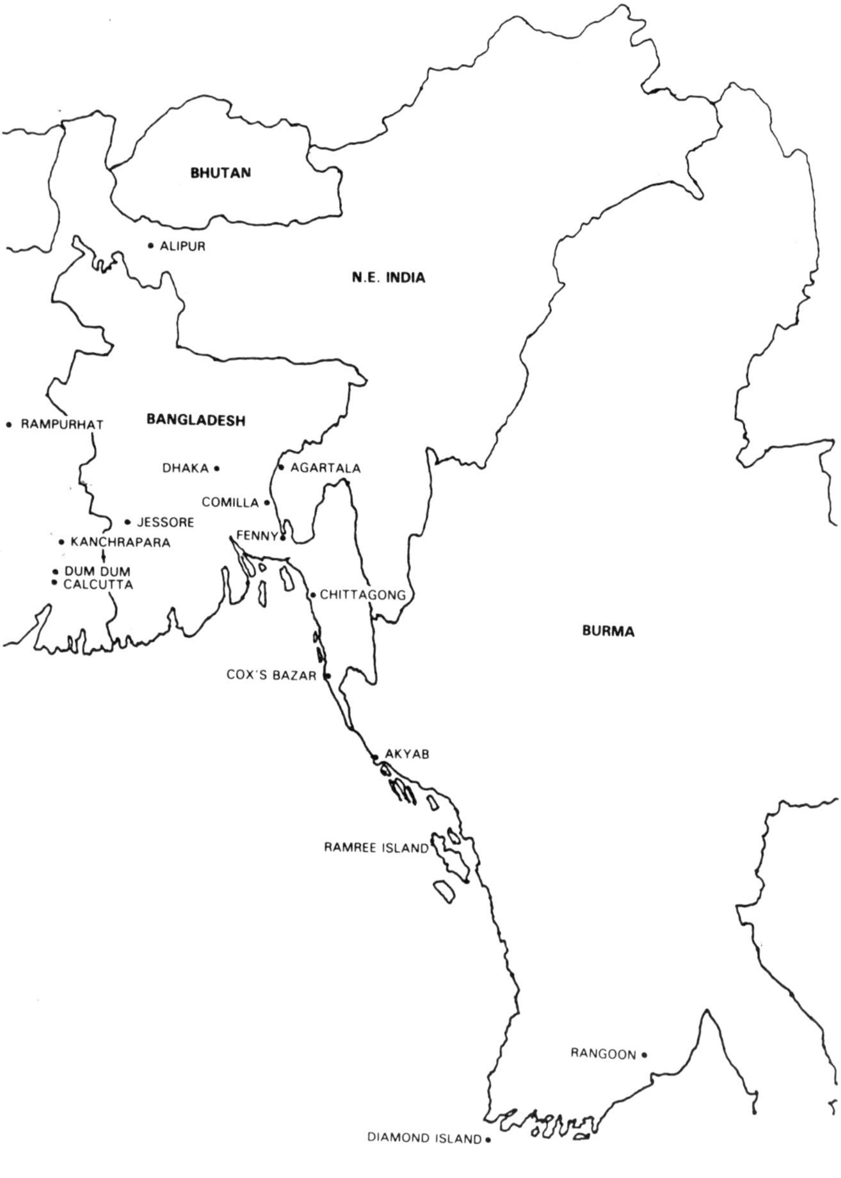


Airfields used in Bengal and Burma

# A Fresh Start — with Fresson

On 2nd January 1946 Eric took de Havilland Rapide G-ADAJ, 'Inverness', up for a twenty-five minute familiarisation flight around Inverness, not having flown a Rapide for six years. On the following day he flew his first scheduled service for Scottish Airways Ltd. in Rapide G-AGDG, a four and a quarter hour stint linking Inverness to Sumburgh via Kirkwall and then returning non-stop to Inverness.

Many of Eric's contemporaries from the services were joining Imperial Airways. But Eric had spent the last five years away from home and his young family, flying long distance patrols across the seas. The thought of joining the inter-continental airline, with long spells away, did not hold out the glamourous prospects that may well have been the case at another time. Eric also had some misgivings about rejoining Allied Airways at Aberdeen. During the war years the various airlines operating internal routes had maintained limited services by pooling their resources through the Associated Airways Joint Committee. Gandar Dower, ever fiercely independent, had remained outwith the A.A.J.C. but through shortages in aircraft and pilots his airline emerged from the war in a weakened condition.

As it happened, by joining Scottish Airways at their northern base located in Inverness, Eric was joining the other great pioneering tradition of pre-war Scottish civil aviation, Scottish Airways Ltd. being an amalgamation of Northern & Scottish Airways of Renfrew and Highland Airways of Inverness. George Nicholson's Northern & Scottish Airways had pioneered routes to Islay, Tiree, Barra, South Uist, North Uist and Skye while Highland Airways, under Edmund Fresson, had competed vigourously with Eric Gandar Dower on the links to Orkney and Shetland.

Fresson had particularly made his mark on Orkney where he had encouraged local communities to aid him in constructing airstrips on islands of the Orkney group. This had enabled him to launch inter-island services to North Ronaldsay, Sanday, Stronsay, Westray, Rousay and Hoy. It also benefited the islands with quick access to medical attention at the Balfour Hospital in Kirkwall by the introduction of air ambulance services.

Born at Wickford, Essex, on 20th September 1891, Edmund Fresson had always been interested in aviation. His early career took him to China in 1911 attached to the engineering department of a trading company. Continuing his travels to North America, it was in Toronto and Fort Worth, Texas, that he undertook training as an aviator. Back in his native England in 1919 he undertook the first 'official' domestic air mail delivery for the Post Office when, during a rail strike, he carried letters from Skegness to Cambridge in an Avro 504K in return for a tank of scarce fuel.

Fresson was soon to return to China where his career took a strange turn and in 1924 he found himself at Taiyuan in Shansi province contracted to design and build three prototype two-seater biplanes and with the promise of a contract for a further fifty upon the successful completion of these. Import restrictions meant that all possible parts had to be made locally, but the wheels and instruments were imported under the guise of motor car parts while the engines from abroad were alluded to be destined for motor-boats.

The completed prototype successfully took to the air in 1926, but a combination of outside competition and political unrest prevented the lucrative follow-up order coming to fruition. As a result Fresson was soon to find himself returning to England and in 1929, with Lance Rimmer, he set up the North British Aviation Co. Ltd. This inaugurated barn-storming tours of England and Scotland that were soon to take Fresson into what he describes as 'like flying into another land with a population that appeared different in some intangible manner'. The feelings experienced upon his arrival in the Orkney Islands were no doubt accentuated by the touching and prodding that the crowd applied to his Gipsy Moth as they tried to evaluate this strange aerial contraption that had so aroused their curiosity.

The visit germinated an idea that culminated in the formation of Highland Airways Ltd, based at Inverness, in 1933. The first route linked Inverness with Wick and Kirkwall while exploratory flights for further routes took Fresson to points such as Stornoway and remote Fair Isle. On 29th May 1934 Fresson launched the first regular internal Air Mail service when he received the Post Office's pennant prior to flying from Inverness to Kirkwall with over 2,000 letters. As he strived to develop services from Aberdeen he soon came into conflict with Gandar Dower's competing airline and this was particularly high-lighted in the race to inaugurate services to Shetland.

In October 1938 Henry Vallance, one of Gandar Dower's most experienced pilots, had left him to join Fresson at Inverness. So how did Gandar Dower view the recruitment of his former Chief Pilot by his rival? In truth both entrepreneurs had other problems on their minds which were to make the former opponents become friends in adversity.

The war in Europe had finished on 7th May 1945 and this paved the way for a General Election on 5th July. This was won by the Labour Party. The new government had pledged a programme of nationalisation and their plans for aviation were revealed on 1st November 1945.

Initial intentions were for a Scottish Division of the British European Airways Corporation to operate with a substantial degree of autonomy. After all, the pioneers in Scotland had run highly successful operations across difficult terrain and had performed an important role during the war that was still fresh in the minds of the whole nation. But the Managing Director of B.E.A. was Gerard d'Erlanger and he seemed to take an instant dislike to those whose experience of aviation outweighed his own.

Edmund Fresson was given the unsavoury task of removing Allied Airways from his old rival, Gandar Dower, on behalf of B.E.A. in April 1947. George Nicholson, founder of Northern and Scottish Airways Ltd., was to be dismissed from the new Corporation in July 1947. Fresson himself was then to be dismissed in February 1948 along with fellow Scottish Airways Director William Cumming, while David Barclay, who had pioneered airfields throughout the Hebrides, was only given a reprieve from similar treatment at the eleventh hour.

While a disillusioned Fresson set off for Kenya to try and build a new life, Gandar Dower continued to fight tooth and nail for adequate compensation for his airline. He retained a small nucleus of his aircraft. These he operated for a short time under his own name before putting them into storage at Dyce where they remained until 1967. Gandar Dower continued his fight for compensation throughout the life of B.E.A., the last claims finally being settled in 1973. Gandar Dower had entered the political fray in the General Election of 1945 winning Caithness & Sutherland constituency for the Unionists by the slimmest majority of just six votes unseating the Liberal incumbent of twenty-two years standing, Sir Archibald Sinclair. But typical of his individualistic style, Gandar Dower broke with the Unionist Party in 1948 and served the remainder of his term in Parliament as an Independent.

Eric started flying a familiar pattern of routes from Inverness in the de Havilland Rapides of Scottish Airways. With the wheel of the familiar bi-plane in his hands and the comforting drone of the reliable 200 h.p. Gypsy IV engines either side of him, the intervening war years had merely been an interruption to the more stable friendship offered by the reliable Rapide for which time had stood still. The route structure linked the Highland capital with Wick, Kirkwall, Sumburgh and Stornoway.

Stornoway had long eluded the airlines despite a decade of spasmodic exploratory flights. However the war years, a period when development of civil air services ground to a halt, witnessed a notable exception when the service to the Isle of Lewis was inaugurated from Inverness on 24th May 1944.

Since the Rapide did not have de-icing equipment the flying ceiling was dictated by the freezing level. By following the valleys when flying from Inverness to Stornoway it was possible to complete the journey without flying higher than 700 feet. Later when the route was operated by B.E.A.'s Dakotas, Tony Millward, the B.E.A. Chairman, considered it a special treat to be flown by the valley route when accompanying Eric from Inverness to Stornoway. A plausible explanation had to be worked out in advance in case any passenger enquired why they were flying with the mountains towering on either side of the Dakota. It didn't end there as Eric narrates.

'I remember flying Scott Hill, one of the head office types, in a Viscount on a sight-seeing flight down the centre of Lewis. As it was a calm, sunny day I flew low, following roads and valleys. Scott Hill accused me of flying the Viscount like a Rapide — with a certain amount of justification.

'The Wick-Kirkwall trip took fifteen minutes in a Viscount. A radio controlled let-down through cloud at Kirkwall took at least a further fifteen minutes. This meant that one doubled the length of the flight if one climbed into cloud. Therefore we ex-Rapide types preferred to creep along under low cloud if at all possible. I was amused one day when I was creeping along at about 300 feet over Scapa Flow, inbound to Kirkwall, when a Coastal Command Shackleton passed below me going the other way.'

On 7th February 1946 Eric performed his first post-war ambulance flight when he carried a stretcher patient from Sumburgh to Renfrew. Strong south-westerly winds hampered the Rapide's progress and a refuelling stop was made at Dyce. The total journey from his Inverness base had taken six hours and thirty minutes flying time. Two weeks later Eric was battling through heavy snow from Inverness to Sumburgh on a scheduled flight. On Shetland visibility was down to 200 yards as he sought out the airfield with diminishing fuel reserves. A few days later, the snow unabated, the second attempt to reach Kirkwall succeeded after the first had been forced back to Inverness.

During 1946 Eric's young son Angus, now six years of age, joined him on several flights taking up an inconspicuous position behind the cockpit.

In May the infamous east coast haar provided some interesting conditions in Orkney and Shetland with one flight to Sumburgh having to be abandoned when visibility deteriorated to 150 yards. However clouds of a different nature were soon to gather at Inverness too. With preparations for the impending nationalisation becoming ever more evident Inverness was no longer to be an operating base. Eric had the choice of moving to Aberdeen or to Renfrew.

With many friends and Eleanor's family connections in Aberdeen, it was an attractive option. However Eric forecast that the days for Aberdeen were also numbered and so chose to move to Renfrew. As it turned out his prophecy on Aberdeen's continuation as a base were soon confirmed. The only aircraft left at Inverness was the ageing Dragon G-ACIT, nicknamed 'Old IT' by those close to her. The wisdom of basing aircraft to serve the northern routes in far-off Renfrew was highly dubious, but was only one of many inappropriate decisions of the time by the new administration.

True to tradition, Eleanor was in the last month of pregnancy when Eric had to move to Renfrew. She had a married sister in Largs so they wired her, 'Get us a furnished house.' This she did and they planned to move nearer to Renfrew after the baby had been born. Elma was born on 28th March 1947, not without drama. Arrangements had been made for Eleanor to have the baby at the maternity hospital in Kilwinning, about twelve miles down the coast. Unfortunately a very heavy snowfall came down. The coast road was impassable and even the railway tunnel at Fairlie was blocked with snow. The thaw came only just in time for Elma to be born in the hospital.

Eric found that commuting from the Ayrshire coast was easier than he had anticipated and petrol was only 10d a gallon. His father had recently died leaving him some money and when a suitable bungalow became available Eric bought it 'cash' for the then enormous sum of £2,450. That house has been the Starling home ever since, nicknamed Hangover Hall by the ever mischievous Eric.

From Renfrew Eric immediately started to fly services to Belfast, Campbeltown, Islay, Tiree, Benbecula and Stornoway while on 23 October 1946 he made his first landing on the famous cockle shell strand that was the airport at Barra. On 25th November his ambulance duties continued with his first west coast operation, between Campbeltown and Renfrew, still termed as an 'Ambulance Charter' since the service had yet to be incorporated into the new National Health Service.

On 30th November 1946 Eric had his first trip on the flight deck of a Dakota when he flew as First Officer on a service from Renfrew to Northolt, the 'London Airport' of the time, and then to Belfast. This aircraft had already made a major impact on operators such as KLM Royal Dutch Airlines and Transcontinental & Western Air (later to become Trans World Airlines) who had introduced this revolutionary machine to the services during the late thirties. The second World War then caused the Douglas DC-3 Dakota to be produced in massive quantities to serve as troop transports. 10,655 of this highly successful aircraft were ultimately built and it now became widely available for civil use. It's two Pratt & Whitney Twin Wasp piston engines were 1,200 h.p. and it could carried 25 passengers.

B.E.A. modified their Dakotas and called their version the Pionair. The most important modification was the fitting of disc brakes. The old shoe type brakes suffered from oil dripping from the engine. Crews seen squirting carbon tetrachloride fire extinguishers on the brakes were not extinguishing fires but were cleaning out the oil. B.E.A. also completely redesigned the dashboard. After several more trips as First Officer, Eric took Dakota G-AGZB up from Renfrew for eighty minutes doing the necessary tests for getting the Dakota endorsed on his licence on 10th January 1947. He commanded his first Dakota flight eight days later on the Renfrew-Northolt run.

On 28th January 1947 Eric did the necessary flying to get the Jupiter on his licence. It was perhaps symbolic of the times for on 1st February 1947 the merger of Scottish Airways was completed and the British European Airways Corporation came into being. The Jupiter arrived in B.E.A. colours, as had the Dakota, in anticiptaion of the impending nationalisation.

'The Jupiter' was the nickname given to Junkers Ju-52 aircraft that had been inherited from Germany as part of war reparations. The Ju-52 had taken to the air in 1931, one of a new breed of metal monoplanes. It had served successfully with many airlines and Eric had experienced them in Norway with D.N.L. when he had toured the country prior to the launching of Allied Airways Aberdeen-Stavanger service in 1937.

The Ju-52s were refurbished by Short & Harland at Belfast, but it was found out too late that they had been mass-produced by Germany during the war. They were crudely made and were intended to fly for about 100 hours before being abandoned on beaches and fields near the theatres of war. On the Scottish internal routes they were costly to operate and frequently became unserviceable throwing schedules into disarray. Service units were required to start the Jupiter's three engines and, as many airfields were not supplied with these, the engines had to be kept running during the loading and unloading of the aircraft. Furthermore, the Jupiters carried fourteen passengers, only three more than the Rapide with its proven reliability and low operating costs.

The Ju-52 had been used by Imperial Airways on services to Scandinavia as a replacement for the de Havilland DH-86 which was subject to icing up. The Ju-52 did not encounter these problems and it was even suggested that ice on the aircraft's corrugated skin provided streamlining and improved its performance during these conditions!

During Scottish Airways' operating life from May 1933 until January 1947, the airline had carried 339,812 passengers and had clocked up nearly 10 million miles. B.E.A. meanwhile was getting off to a slow start. On only the third day of the new corporation Eric recorded that his Renfrew to Northolt flight had flown at 12,000 feet in both directions in order to avoid static snow and that his aircraft was the only one to land at Northolt that morning.

Eric continued to fly a mixture of services; Dakotas to Northolt and Belfast and Rapides to the Western Isles with the Jupiter edging its way in as the year progressed. During March a little variety appeared with a charter to Mull to photograph a shipwreck.

On 27th May 1947 Eric was flying to Orkney from Renfrew via Aberdeen when he encountered reciprocal bearings at Kirkwall. These were bearings that were 180 degrees out and this was the second time that Eric had encountered such an instance. On a previous occasion at Wick, Eric had been given reciprocal bearings which meant he was letting down straight for the Sutherland hills. Fortunately Eric used the old trick of turning the wrong way and if the bearings followed the aircraft they were reciprocals. He realised that he was getting reciprocals and made appropriate remarks over the radio. The situation could have been fatal and Eric promptly phoned the Ministry of Aviation. The wireless officer involved was due to retire within two months and Eric was dissuaded from pursuing the incident since a formal complaint would have deprived the man of his pension. Eric agreed to be persuaded not to take the matter further.

On 1st July 1947 Eric was appointed Flight Captain for Rapides and Dakotas at Renfrew. Eric had his sights on this job, which was about to be vacated, when he had moved to Renfrew from Inverness. However, George Rayer, a pilot senior to Eric, decided to move up from England and so got the appointment. After a year however Eric had persuaded George to move south again! Yet during that month Eric flew only once in the Dakota and not at all in the Rapide.

Instead he was becoming familiarised with a new aircraft type, the Vickers Viking. Following a week of training at Aldermaston he then flew on the aircraft from Northolt to Amsterdam and Copenhagen to gain route experience.

The Warwick undergoes cowling repairs while on the ground at Sharjah.

Office work at Agartala.

The Mozumder Family. Standing: Shiuli, Arthur Hutchings and Gora. Seated: Eric, little Gopa, Luxmi and Moni.

Eric at the controls of his Harvard taken by Pete Almack from a Sea Otter amphibian.

A visit to the shops in Agartala.

Group photo at Jessore before a Sea Otter. The front foursome are Wing Commander Eric Starling, Station Commander of C.O. 292 Squadron; Air Vice Marshall Sir Guy Garrod, Commander in Chief; Wing Commander Hodges, C.O. 'Cloak & Dagger' Squadron; Squadron Leader Hucklebridge, C.O. Maintenance Squadron.

A Dragon Rapide traverses a snow draped landscape during the early years of B.E.A.

Edmund Fresson and the General Aircraft ST4 Monospar with which he inaugurated Highland Airways' first services.
*(Courtesy of Peter Clegg)*

During a proving flight to Copenhagen and Gothenburg several pilots were travelling on the Viking, taking turns to fly the aircraft. While Eric was at the controls during the return journey he felt that the aircraft was a bit unstable. His wireless operator opened the flight deck door and an arrow came flying past Eric's head, emplanting itself firmly on the cockpit window with the sucker at its tip. Eric had bought the toy bow and arrow for his son, Angus, while they were in Copenhagen but his colleagues in the cabin had decided to try it out by competing with their skills for the Danish coins that they all had left over after the visit. The target had been the notice on 'How to use a Life Jacket' on the cockpit door — until the wireless operator had opened it with such precise timing.

On 23rd July 1947 Eric flew Viking G-AHPA 'Varlet' from Northolt to Renfrew. The first visit of the Viking to Glasgow brought with it an incredible new record, completing the proving flight in only one hour and twenty five minutes, comparable to the scheduled time by modern jet aircraft over forty years later. The normal time for the 348 mile trip by Dakota was two hours while the Viking would have a scheduled time of 1 hour 55 minutes. This fast time had been made possible because Eric was able to turn straight on course after take-off and aim straight for Renfrew. Later special inbound and outbound routes were introduced so that aircraft descending into the airport didn't collide with the outbound ones climbing on to course.

One of the most unpleasant cases of this Eric experienced when he was going into Brussels one night. Just as he reached the holding beacon with several aircraft at 1,000 feet intervals below him, the ground radio station broke down. What did one do? Diverting was dangerous because of incoming aircraft. Then an Aer Lingus pilot called, 'I'm the bottom of the pack and am going in so keep out of the way'. Shortly after that Brussels Tower came on the air again but not very strongly. It transpired that they had enterprisingly resumed by controlling from an aircraft on the tarmac.

If Eric were asked, 'What's your favourite aeroplane?' he would usually answer, 'The one I'm flying at the moment!' However he did not take a liking to the Viking at all despite his record-breaking flight. It earned little respect from him.

'It was a horrible aeroplane. It was a sort of converted Wellington and while a Wellington was an ideal military aeroplane, it was useless as a civil aircraft. Every time you put your wheels down, changed your flap setting, or changed the throttle setting, the trim of the Viking changed and it was incredibly difficult to keep on a glide path when you were doing a beam landing.'

The Vikings carried a series of incongruous names such as Vimy, Vindictive, Vernal and Verity, the first letter being the only semblance of a theme. One of their number, G-AIVE 'Vestal', crashed during its approach to Renfrew on 21st April 1948, fortunately without loss of life. In 1949 the first Vikings had the fuselage extended and tailplane enlarged and these modifications improved their stability. At the same time they were given new names so that the type reflected a more appropriate nautical theme which honoured famous Admirals.

The Friday evening flight from Northolt was always very full with businessmen and M.P.s returning home. Eric was taking a Viking north on the last flight. He had been held up signing the ship's papers and followed the passengers on board. As it was the last service and the traffic staff wanted to get away, they slammed the door behind Eric, pulled the steps away and quickly disappeared.

Eric had to wait until all the passengers were seated before he could get to the front but he found that someone was still standing after all the seats had been filled. He discovered that the extra passenger was a stowaway Scottish stewardess who had sneaked aboard. To have her removed would cause a long delay so Eric told her to lose herself in the toilet. Arriving on the flight deck he found Bill Baillie, the Chief Flight Manager sitting on the 'jump seat'. 'I hope you don't mind,' he said, 'I knew that the aircraft was full and I must get back to Scotland tonight.' 'If you don't mind being two over the top, it's alright by me,' responded Eric. A long time later he admitted to Eric that he had been deeply concerned since they would both have been in serious trouble had anything gone wrong.

The introduction of the Vikings released Dakotas for use further north and in August 1947 Eric flew a proving flight to Orkney via Dalcross and Lossiemouth. The replacement of Longman Airfield by Dalcross to serve Inverness had been the latest development to put Edmund Fresson on a collision course with the B.E.A. hierarchy in London. He rightly argued that Longman had low running costs and waiting facilities for passengers, now more important than ever with Jupiters breaking down all over the network, but he was fighting a power much greater than himself.

Angus joined his father on another Dakota proving flight from Renfrew to Stornoway via Tiree and Benbecula. Then on 9th August Fresson won his only battle with London when Eric flew the last Jupiter flight Renfrew-Inverness-Wick-Kirkwall-Sumburgh-Wick-Aberdeen-Renfrew. The final decision to bring about the withdrawal of the Jupiter had not been running and maintenance costs, but the aircraft's risk as a health hazard after a crew had barely got an aircraft down while being asphyxiated by carbon monoxide gas from the engines. The Jupiter had survived for less than six months.

When B.E.A. decided to scrap the Jupiters a lot of spare corrugated Ju-52 sheeting accumulated in the stores at Renfrew. George Griffiths, the Chief Engineer, gave Eric this sheeting which he used to line his loft, a purpose which it still serves to this day.

That more modern machines were appearing on the horizon was illustrated when a Sikorsky S-51 helicopter visited Renfrew on 29th September 1947. Its pilot, Mr Theilmann, took Eric and two colleagues up for a five minute circuit in this strange flying machine. The pilot also demonstrated circuits and splashes in the sea off Prestwick — the Sikorsky was fitted with floats. On the way back, Eric introduced Mr Theilmann to an old R.A.F. trick used with the Air Sea Rescue Sea Otters by suggesting a landing in one of the reservoirs to wash off the corrosive salt water.

Eric concluded 1947 flying regular services with the Dakota with an occasional deviation on to a Rapide or the Viking. One Dakota flight bound for Belfast however didn't get very far. Eric was back on the ground looking for a replacement aircraft after being airborne for only five minutes. A golden plover that had not being watching where it was flying had broken his windscreen.

On 24th January 1948 Eric took de Havilland Dragon DH-84 G-ACIT up for a trial flight after it had been overhauled. With Edmund Fresson's departure imminent, he had sought permission to buy 'Old IT' but this had been denied by B.E.A. Just to make their response totally emphatic 'Old IT' had promptly and furtively been removed from Inverness to Renfrew.

Before Edmund Fresson parted company with B.E.A. he did take pleasure in performing one last duty on 16th February 1948 when he officiated at Dalcross for the first Air Mail service between Inverness and Stornoway. Assistant Postmaster John Brown presented Fresson with the Royal Mail pennant and handed the first bag of mail to Eric. Eric then flew Dakota G-AHCV to Stornoway with Fresson joining the flight as a passenger. By 31st March 1948 Fresson had worked out his redundancy notice and Scotland had lost one of its most notable aviation pioneers.

BRITISH EUROPEAN AIRWAYS

*Telegrams* BEALINE Northolt *Telephone* Ruislip 6061

HEADQUARTERS
NORTHOLT AIRPORT
RUISLIP MIDDLESEX

SF.564

13th January, 1947.

Captain E.A. Starling,
c/o. Scottish Airways,
Renfrew,
Scotland.

Dear Sir,

I have pleasure in offering you employment with British European Airways on the following terms, with effect from the date on which control of your present Company is vested in B.E.A.

You will be eligible for transfer in the rank of Junior Captain at an annual salary of £ 900- and a Cost of Living Allowance of £ 78- per annum.

The incremental date applicable in your present employment will remain unchanged.

You will be required to enter into membership of the Pension Scheme applicable to your case, in accordance with the rules governing such Scheme which is being negotiated in consultation with the Minister of Civil Aviation and the British Air Line Pilots Association.

This appointment is subject to termination by either party giving to the other not less than three calendar months previous notice in writing.

With effect from the date of your employment by B.E.A., you will be insured against personal accident risks for the sum of £4,000- the premium for this insurance being payable by B.E.A.

Your employment will be governed by B.E.A. Orders and Instructions at present in force, and as amended from time to time. You will be deemed to be

/over

aware of all such Orders and Instructions, which can be obtained from past issues. Current Orders and Instructions are displayed from time to time on notice boards.

Please indicate whether you accept this appointment in the space provided at the foot of this letter and return it as early as possible to this Office. The second copy of this letter should be retained by you for reference.

Yours faithfully,
for StaffManager.

S.P. Hagger.
Flying Staff Superintendent.
British European Airways Corporation.

From:

I accept/do not accept employment with British European Airways Corporation on the terms contained above.

Date: Signature:

# In Scottish Skies

As the nineteen fifties dawned, B.E.A. was operating its main Scottish services with the DC-3 Dakota. These were in fact Douglas C-47As of war-time vintage and B.E.A. had twenty-six of them in its fleet. On thinner routes the De Havilland Rapide still continued to give the sterling service that it had offered in the thirties and which the Jupiters had failed to usurp from them.

On 1st April 1949 Eric flew Dakota G-AJIS on the inaugural service from Edinburgh's Turnhouse Airport to London's Northolt Airport. The three hour flight commenced from Renfrew. It was the pattern from this time for the first Edinburgh-London flight to start from Renfrew and the last northbound flight of the day would likewise continue from Edinburgh to Renfrew since the maintenance base was there. As a result quite a few passengers from Glasgow would use these flights for an aerial day trip to Edinburgh.

On one occasion Robert McKean, General Manager of B.E.A. Scottish Division, was a passenger on the Edinburgh-Renfrew flight. Eric let the First Officer do the landing at Renfrew and the aircraft came down with a thump. Robert McKean commented on the lack of finesse displayed by the touchdown but as 'staff' he had only paid 10% of what was already a very low fare. Eric was quick to respond, 'You don't expect a Captain's landing for 1/6d, do you?'

The Dakotas had a small hold on the port side of the flight deck with a loading door which opened just in front of the port engine. The cabin heating was often unreliable and flying to London one cold winter morning Eric discovered, when he went back to talk to the passengers, that they were shivering with cold. The heating was fine in the front so, as there were only four of them, Eric suggested to the passengers that they might prefer to move forward and sit on their cases in the front freight hold. They agreed to this but Eric often wondered what investigators may have thought if the aircraft had crashed and everyone was found in the flight deck.

In 1950 it was decided that the Douglas C-47A Dakotas were no longer of a satisfactory standard for passenger services. A programme of refurbishment was instigated and, one by one, they were rebuilt and converted to a developed version of the civil Douglas DC-3. The remodelled aircraft had more modern radio and instruments enabling two-crew operation instead of three-crew operation. The passengers enjoyed greatly improved amenities and comfort with innovations such as reliable cabin heating and built-in steps at the door. The conversion work was undertaken by Scottish Aviation at Prestwick. However, when the prototype was completed in September 1950, it was almost a totally new aircraft that emerged and so B.E.A. called it the 'Pionair'. Eventually 38 Pionairs were in service with the airline.

Eric flew a proving flight from Renfrew to Perth on 30th June 1950 and he captained Rapide G-AGUR for the inaugural service the next day. On board were Sir Patrick Dollan and

Robert McKean. Although McKean is not recorded as being unhappy with the landing on this occasion, the Renfrew-Perth service did not continue for long. It was the only time that Eric flew the route during B.E.A. service.

There was still room for a little informality to slip in from time to time and this was particularly true of island flying. For example on 15th August 1954 Eric recorded meeting a golden eagle high above North Uist and he proceeded to circle it. Eric was captivated by the magnificence of this mighty bird, and the golden eagle no doubt viewed Eric's Rapide with similar wonder.

In 1954 Eric was already celebrating anniversaries. On 4th September he flew into Campbeltown with Dakota G-ALXK to be greeted by James McGeachy with a presentation box containing some white heather picked from Island Davaar at the mouth of Campbeltown Loch. Eric was delighted by the gesture and that McGeachy had remembered that it was twenty years ago on that day that he had first landed at Campbeltown with Gander Dower and John Sword when Aberdeen Airways had been contemplating taking over the operations of Midland and Scottish Air Ferries.

The service to Campbeltown always continued on to Islay where the Station Superintendent found that the peat cutting that went with his house was producing more dross than he knew what to do with since peat only burns satisfactorily in lumps. Eric suggested that he put it in bags and send it to him at Renfrew whenever there was space on the daily service. He was never short of garden peat while that arrangement lasted. It reminded Eric of a similar arrangement that he had learned of in the United States. A pilot had discovered at one of the places where he night-stopped that bankrupt stock was being disposed of at a monumental mason's yard. Blank gravestones were being sold at knock-down prices and he decided that these would be ideal for paving the driveway to his garage. A consignment was delivered to the cargo section of his airline where it was arranged that they would be gradually shipped to him on any aircraft with some spare capacity. When they reached his base they were stacked outside the cargo hangar until there was sufficient for him to acquire a truck to shift them. The system worked fine but a salesman from a rival airline was watching the regular movements of headstones, decided that he was missing out on some lucrative freighting business, and persuaded his office to publicise a special cut rate for gravestones. The pilot's own airline responded by offering an even cheaper rate. Neither could understand why this price war produced absolutely no results.

Donald Macdonald was Station Superintendent at Benbecula for many years. It was not until well after the Second World War that Benbecula was linked to North Uist by causeways. Until then people had to cross at low tide in a horse and cart, the driver of which was the only person who knew how to circumnavigate the dangerous quicksands between the two islands. Donald reckoned the risks were greatly exaggerated for business reasons since R.A.F. transport was crossing continuously during the war and he only knew of one vehicle being lost in the quicksands.

One day Eric arrived at Benbecula with a very smooth-talking door-to-door salesman. The salesman had arranged for the horse-drawn 'ferry' to meet him at the airport but there was no horse and cart awaiting him when he stepped from the aircraft, probably because the tide hadn't receded sufficiently for it to cross from North Uist. The salesman unleashed his frustration on Donald who had tried to explain the probable reason for the delay. The salesman stormed off the airfield and Eric, who had been observing the exchange, enquired where he had gone. 'I don't know,' said Donald. 'There's nowhere that he can go. He'll be back soon and he'll probably listen to me this time.'

One crofter who lived right off the beaten track used to take advantage of a low postal subscription rate for newspapers and the postman had to make a long detour every morning to deliver his Scottish Daily Express. The postman had to leave his van by the road to trudge down a long muddy track with the paper. After a time the wily old crofter would say to the postman that, as he would be back again tomorrow, if he gave him some money could he bring him some butter or some other provision. The postman was soon doing his shopping as well as delivering his paper.

One day Eric had one of the senior Post Office staff on board. He was making a visit to the Postmaster in Kirkwall. During his visit the Postmaster asked him if he would care to join him on a visit to one of the Northern Isles of Orkney. The dear old lady who ran the only shop there also ran the Post Office. She didn't take the job very seriously and drove the Postmaster to despair by sending him incomplete returns and books with unfilled-in duplicates. Perhaps the official from Edinburgh could use his weight to enforce the importance of accurate book-keeping. The old lady listened politely to what he had to say then turned around, unearthed a cigar box from under a pile of papers and handed it to him with 'I'm fed up with your wretched Post Office. Take it!'. Both men retreated suitably reduced in stature.

Perhaps the Orcadians rightly felt that they had been dealing with the Post Office long enough to know when they could get the better of it. On 31st May 1955 Eric flew newly delivered Heron G-ANXA from Renfrew to Kirkwall via Inverness to mark the anniversary of the first internal Air Mail flight on the Inverness-Kirkwall sector twenty-one years earlier.

It took more than a bit of authority to impress islanders. When Lord Douglas of Kirtleside, Marshall of the Air Force, became Chairman of B.E.A., he decided to do a trip round the Scottish islands in one of the Rapides. In due course he arrived at Tiree where he was greeted by a completely empty tarmac. As Marshall of the Air Force he had become accustomed to a bit of a parade when doing his rounds. Eventually Colin MacPhail, the Station Superintendent, appeared from a building, probably wearing a creased uniform and no cap. When he opened the door of the Rapide, the Air Marshall roared at him, 'Weren't you expecting me?' Colin, in his soft Hebridean voice calmly replied, 'If you're Lord Douglas, I was. If not, I wasn't!'

At Barra's beach airfield Katie Macpherson had a walkie-talkie which she used to advise the pilots of the landing conditions. On 4th October 1958 the system must have failed since Eric recorded bending the flaps on his Heron when landing in deep water. The rule of thumb used to be that if you could see the seagulls' knees the water wasn't too deep for a landing.

The people of Lewis are strict observers of the Sabbath and everything comes to a standstill on the island on a Sunday. While it was greatly frowned upon by the islanders, occasionally a flight to Stornoway had to be operated on Sunday if there had been cancellations during the week and a back-log of passengers and cargo had resulted. Such an incidence had occurred on Saturday 17th July 1954 when the flight had been cancelled because of bad weather, (the weather is traditionally atrocious at the Glasgow Fair, the Glasgow trades holiday which takes place during the second fortnight of July).

There were a large number of passengers so it was decided to operate a service the following day which Eric volunteered to fly. He took off in a Dakota but had only been flying for ten minutes, gaining height over the Clyde, when one of the engines began to give trouble. The propeller had to be feathered and Eric had to return to Glasgow and make a landing on one engine with his full load of passengers. Another Dakota was substituted and off they went again to Stornoway after a call at Benbecula. On final approach to Stornoway there was a terrific clatter which sounded like something striking the flaps. Eric landed the Dakota but when he started to taxi one of the engines made a noise like a traction engine. Eric didn't want to stop on the runway and have to tell his passengers that he had a second engine failure so he continued to taxi as best he could. The engine produced the most horrible grinding noises which Eric hoped were not too audible in the passenger cabin. It transpired that in one of the cylinders, the piston had come adrift from the conrod. After that there was no shortage of moralisers around telling Eric, 'That will teach you to fly to Stornoway on the Sabbath!'

Some years later, on 2nd September 1965, Eric was flying a Viscount to Benbecula with a Training Captain observing him for his six monthly route check. He was approaching Benbecula where the weather was 'on-limits'. He worked his way down to the critical limit for Benbecula of 300 feet and the ground was still nowhere to be seen. An overshoot was then required and Eric was just starting to do this, under test conditions with the critical eye of the Training Captain on him, when the Steward came into the cockpit saying, 'There's a passenger in the back. I think he has died'. Eric managed to keep his composure despite being on the receiving end of this dramatic news. The overshoot was executed and they continued on to Stornoway where, sure enough, the unfortunate passenger had passed away from natural causes during the course of the flight.

Characters were still tolerated during the early years of post-war civil flying. One of these was Johnny Welford, a flamboyant James Robertson Justice kind of character who was sadly killed in a car accident only two weeks before he was due to retire. When he joined Scottish Airways after the war he had to take up his flying duties before his uniform had been made. Being the man that he was, he bought himself a heavily checked tweed jacket and a deer stalker hat which he wore when he was flying.

One day he arrived at his Rapide to find that the passengers were already seated and waiting for him. Without hesitation, he sat himself down in an empty passenger seat and began complaining that Scottish Airways pilots were always late. After keeping this up for a while, he told all and sundry that if the pilot didn't turn up soon he would fly it himself because he had an urgent appointment in Stornoway. After an interval he said he wasn't waiting any longer, shut the door, climbed into the front, started the engines and off he went. This story has been appended to travellers experiences in many parts of the world, and perhaps other cavalier pilots also pulled this stunt in the old days. I am assured that Johnny Welford certainly did it at least once and he may indeed have been the originator of the prank.

Some years later Johnny was still up to mischief when he greeted the passengers at the top of the steps of a B.E.A. Viscount at London Airport. He welcomed the passengers aboard with open arms and told them to come up to the flight deck as often as they liked since they would be more than welcome. When the last passenger had boarded, he went down the steps and walked over to another Viscount — the one that he was flying.

In the days when aircraft were not so fast and there was time for the Captain to go back and talk to the passengers, Captain Panda Watson would, on say a flight to Paris, enter the passenger compartment and announce to anyone within hearing, 'Isn't this a lovely day for the flight to Rome.' Two of Eric's nieces were travelling on one of Watson's flights when he nonchalantly gained the instant attention of his passengers with this ruse. More recently the author was on a flight from Glasgow to Stornoway when the First Officer's voice came through the address system of the BAe 748 to express his wish that the passengers were enjoying a relaxed flight to Benbecula. The Captain's voice quickly followed to inform the now unrelaxed passengers that he assumed no responsibility for the First Officer's navigational abilities.

Passengers could prove interesting as much for their importance as for the anecdotes that could be ascribed to them. Eric had his share of V.I.P.'s on the Scottish routes.

On 15th May 1951 Eric flew a Rapide from Renfrew to Campbeltown carrying the Australian Prime Minister, Sir Robert Menzies.

Having safely delivered the Australian Premier Eric was entrusted with members of the Royal Family on several occasions. During the course of Scottish operations Eric few Prince Charles from Inverness to Renfrew on 26th July 1962. A student at Gordonstoun, Charles was one of three young princes on board Heron G-ANXA, the others being Prince Alexander of Yugoslavia and Prince Welf of Saxony.

On 25th August of the same year Eric had Princess Margaret and Lord Snowdon travel on board his aircraft, Viscount G-AMOP, from Edinburgh to Aberdeen, a low key affair as they were travelling incognito.

Eric had a previous brush with the Royal Family on 9th November 1953. The Royal Yacht 'Britannia' was undergoing trials on the Clyde. A newspaper chartered a Rapide so that their photographers could record the ship's progress. They thought it was to be an added bonus when the Duke of Edinburgh, who had joined the vessel for the trials, appeared on the bridge. However the Duke could obviously read the minds of the press men. Whenever Eric flew down one side of the ship Prince Philip walked to the other side of the bridge. The photographers didn't succeed in catching him with their cameras but Eric certainly kept the Duke walking to and fro.

Of the showbiz personalities, two regular users of the Glasgow-London service elicited totally opposite sentiments from the traffic staff at Renfrew. Bob Hope often flew into Renfrew en route to Gleneagles for a golfing weekend. If the taxi journey back from Gleneagles on the Monday was slightly delayed by traffic, the London service was held back so that he could catch the flight. He was always so appreciative when things like this were done for him that the traffic staff had nothing but praise and affection for him. However the late Gilbert Harding used to be so objectionable polishing his ego on helpless traffic staff that it was seriously considered that he should be banished from B.E.A. services. The airline was terrified that one day a traffic clerk would feel unable to take any more of the Harding treatment and might respond in kind. As the customer is always right, B.E.A. would then regrettably have had to take Harding's side. Other showbiz personalities included Clement Freud on an Islay flight, while the cast of 'Rocket's Galore', led by the immortal Duncan MacRae, were flown to Barra during the making of this classic film of Hebridean insurrection against Whitehall bureaucracy and contempt.

A special flight took place during December 1961. The passengers were all senior B.E.A. personnel and other aviation officials. But the V.I.P.s were on the ground — the members of the communities all around Scotland to whom the airline was the link with the rest of the country. The aircraft was G-APWA, the first production model of the Handley Page Herald, and it was on a demonstration tour of the islands where it was soon to be introduced as a replacement for the Dakotas. The forty-four passenger aircraft visited Renfrew, Islay, Campbeltown, Sumburgh, Kirkwall, Aberdeen, Inverness and Edinburgh. The first Heralds arrived at Renfrew the following March and Eric did the appropriate flying to have it added to his licence.

On 2nd May 1966 services at Glasgow were transferred from Renfrew to Abbotsinch with the opening of the new Glasgow Airport. The airport had been designed by Sir Basil Spence at Edinburgh. Eric's son Angus was now an architect with Spence and Margaret, Angus's future wife, was a secretary with the company. At 5.36am that morning Eric flew Viscount G-AOYP from Renfrew to Edinburgh to pick up the full complement of Basil Spence's staff and fly them to Abbotsinch on the first passenger flight. Departure from Turnhouse was 6.33am, with arrival at Abbotsinch at 7.00am followed by an official breakfast. The Control Tower assured Eric that he would be the first flight to land. 'If anyone else comes in, we'll stand them off until you arrive!'

On Friday 8th October 1971 Eric had rostered himself to fly the Glasgow-Tiree-Barra service. He wanted Eleanor to join him on this flight and arranged for a staff ticket for her. This ticket was on a stand-by basis but he asked the ticket office staff to do him a special favour by ensuring that a seat would be available for her. 'What's it all about?' he was asked. Eric then revealed that it was the fortieth anniversary of his first solo flight. 'That's alright,' was the response, 'We'll give her a free ticket — provided you don't mind us telling the Press.' Things then got a little out of hand, although the rumour that was put about suggesting that there were so many Press on board that a second flight had to be operated for the regular passengers was a bit of an exaggeration.

When the special day dawned it was accompanied by beautiful flying conditions and on arrival at Barra, Katie Macpherson, the island's legendary Station Superintendent, had sherry ready to serve to the passengers and other guests such as the local doctor and district nurse. As Eric was flying, he couldn't drink, but that was rectified when he got back to Glasgow where B.E.A. had laid on a reception.

When a situation arose to which there wasn't a handbook solution, Eric could always be relied upon to come up with some kind of suggestion. In the days of the de Havilland Dragons it had been the custom to recruit some heavy bodies to hang on to the wingtips in high winds to prevent the aircraft being blown over. When Dakotas were being flown to the islands they were too large for assistance by ground parties but could still succumb to strong winds. Eric issued instructions that, during gale conditions, the Dakotas were not to taxi to the terminal but should change passengers on the runway. The fire tender should park in front of the aircraft to break the airflow which was trying to lift it from the ground. When ready for take-off, the crew were to release the brakes and let the aircraft be blown gently backwards to the beginning of the runway. The instruction was probably unique in its unorthodoxy but it put an end to aircraft becoming stuck in the mud at the side of the taxiways because the pilot had been unable to control it while taxying.

On a Dakota flight from Renfrew to Belfast, Eric had Two Ton Tessie O'Shea as one of his passengers. She remarked to Eric that she could never strap herself into her seat because the belts were always too short. Eric passed the comment on to head office and all aircraft carried extension belts from that time on.

The first Dakotas to fly from Renfrew were eighteen-seaters that had been inherited from Imperial Airways. They didn't have a stopwatch but this was required when doing a let down (referred to as 'slices of cake'). The aircraft would fly out for two minutes doing a steady turn to the left, then fly back for two minutes, letting down continuously. There was about a minute on the turn so the altitude was divided by five minutes and eventually the aircraft came out of the cloud with the runway ahead. Eric was unwilling to buy a stopwatch so he settled for a two-minute egg-timer. He used to get his leg pulled but when the sand ran out he would turn the aircraft and flick over the egg-timer. It worked perfectly.

Eric's tricks were not limited to the air. Captain Iain Crosbie recalls being given a lift home to Lochwinnoch by Eric in his Ford Anglia one foggy night. Although it was a saloon, the top half of the windscreen folded down and by sitting on his case Eric could look through the open gap and drive at the same time. It was a trick which had previous applications. When he was looking for fields for the Flying Fair, Eric was driving an old Austin tourer when he ran into similar conditions and applied the same solution.

'Windscreen wipers, heaters and sprayers didn't exist so I just sat on my suitcase and looked over the top of the windscreen. This seems to have become a common practice of mine. The Wellington roof could be opened up above the flight deck. I have long legs, so sitting on the back of the seat with the top open, I could still work the rudders. Doing this one could see right round the Wellington, invaluable when taxying in congested places. This reminds me that the Junkers Ju-52 also had a roof that could be opened above the flight deck. Before the War, the Norwegians flew Ju-52s on floats. When I flew with them, I was amused to see that when taxying on the water they opened the roof, stood right up between the two pilots seats and, looking outside, manoeuvred the aircraft with the throttles which were then conveniently at waist level. Perhaps it was this which inspired me to do something similar in the Wellingtons.'

Eric's creative mind combined with his man-management skills found a non-aviation application when he decided that his old friend Craig Miller needed a bit of a boost. Miller had been a Chief Designer with Rootes and he designed the Hillman Imp. Later he lost both legs through bad circulation and Eric found him after he had returned from hospital lying flat on his back, unable to sit up and having given up what have must have seemed a very unfair struggle. Eric suggested that, if he made him a shiny board, his wife with help would be able to slide him upright on to a wheelchair. Eric got Craig to design the board and thus gained his interest. This got him going and soon they were making regular visits to the local pub together. Later he was provided with a kind of crane which his wife could use unaided to lift him on and off the chair. He was heavy and the crane was a bit insecure so Eric adapted it so that it could be anchored to the wall.

When not creating his own innovations, Eric took a keen interest in others that were taking place in the industry. He therefore made sure that he was on hand when the experimental Bristol Brabazon airliner paid a visit to Prestwick on 29th August 1951. At that time aircraft plying between London and New York had to stage their journey through such points as Shannon, Reykjavik, Bluie West, Goose Bay and Gander. The Brabazon had been designed to carry one hundred passengers in sumptuous comfort non-stop between the two cities and first took to the air from Filton on 4th September 1949. Its multi-decked fuselage included beds for First Class passengers, Lounge and Cocktail Bar and a Ladies Powder Room. Its eight engines were housed in the wings which had a span of 70.10 meters; the wingspan of the latest Boeing 747 400 series is 64.67 meters. The Brabazon is still the largest ever British aircraft. Eric recalls his flight as follows.

'The Brabazon was coming to Prestwick to do a demonstration flight and I phoned Jimmy Jeffs, the Controller at Prestwick, to see if there was any chance of getting on board. I was duly invited to join the demonstration and spent practically the whole trip in the sharp end. It was an unusual aircraft in that the cockpit contained just the basic flight instruments, the blind-flying panel by the captain and co-pilot, and four throttles in the middle. Sitting behind was the Flight Engineer who did everything, literally. When it was time to start, the Captain turned to the Flight Engineer and said, 'Will you start up the engines,' (all eight of them). Then, when he was ready, the Captain said, 'Will you give me the throttles.' The throttles were coupled up in pairs.

'Off we went but, when we were coming into land, the intercom broke down and we all had to shout at each other, the Captain shouting to the Flight Engineer to lower the wheels and put the flaps down. Then as we landed we went right through a flock of starlings which came splattering off the windscreen. The engines were in great big holes in the wings and, after we had landed, a fellow climbed up and proceeded to throw dead starlings out. During the flight I had gone aft where one cabin had been furnished. It was a strange square cabin with a cocktail bar in one corner where other guests were standing around drinking. There were a few rows of seats which, to me, took on the appearance of one of the small news cinemas which were quite common at that time. Some guests had been sitting when, without warning, the carpet moved, a trap-door opened and an engineer emerged and walked off, quite unaware of the surprise that his entrance had created.'

The Brabazon didn't go into regular production and service as B.O.A.C. didn't believe that they could operate it profitably and therefore declined to commit themselves to it. Eric had doubts over the strength of the aircraft as it had gust detectors on board which altered the controls to minimise the strain. 'I didn't like the sound of those,' he concludes.

With the arrival of the Viscounts, Eric still had new innovations up his sleeve, one of which he discovered had earned him the nickname of 'Kipper Feet' at Aberdeen. This came about because the Viscount's inner propellors were very near to the ground and if the undercarriage oleos were a bit soft it was possible to clip the occasional minor obstruction at the edge of the taxiway. Eric found that his Size 12 shoes were perfect for checking their ground clearance.

If the ground clearance was less than his shoe length, he instructed the engineering team to pump up the oleos. The sight of him doing this unorthodox but effective check, together with his characteristic way of walking, inspired the nickname. But Eric's ideas for improving Viscount operation were not limited to his individual way of checking if the tyres needed pumping up.

For many years altimeters had been designed with three needles and the altimeter face was divided from zero to ten. The big needle went round once every 100 feet, the middle needle went round every 1,000 feet and the small needle rotated every 10,000 feet. This worked fine in the days of unpressurised aircraft because they rarely went above 10,000 feet.

With the arrival of the gas turbine engine, pilots started flying at 20,000 feet upwards and confusion started to arise with the reading of the three-needle altimeter.

Someone arrived at London Airport and was cleared down to 5,000 feet. Having descended to that height, he couldn't understand why he was still in cloud when the Met Office had reported no cloud below 8,000 feet. It then transpired that he was still 10,000 feet higher than he should have been. A more serious case arose in B.E.A. when a pilot misread his altimeter 10,000 feet the wrong way and hit the ground near Prestwick, fortunately with no fatalities.

Designing a new altimeter, getting it manufactured, approved and fitted to a fleet of aircraft takes a long time. Eric thought that there must be a way of modifying the existing altimeters as an interim measure. He gave the subject much thought and soon submitted a suggested modification which would provide a 'quick-fix' short-term solution. This included the incorporation of a white arc to indicate the 10,000 foot range. Smiths Instruments made a prototype from Eric's plan and this was installed in one of the B.E.A. simulators for pilot assessment. Everyone liked it and so all the B.E.A. altimeters were modified accordingly while a completely new altimeter was being designed.

Eric thought that, with the modification work completed, the story had ended. However, unknown to him, a Swiss magazine had published an article on the Starling Altimeter and had mentioned Eric as having instigated its design. Also, unknown to Eric, the Americans were having exactly the same problems with altimeter misreadings.

Much later, after B.E.A. had already installed newly designed altimeters in place of Eric's modified version, he was bowled over when he received a letter from Captain Edmund Milauckas, a Human Performance Engineering Officer of the United States Air Force. Milauckas had been given the job of assessing every suggested improvement to the existing altimeters. He had come across Eric's modification design in the Swiss magazine and it proved to be the best under rigorous competitive testing. The U.S.A.F. subsequently used Eric's modification whenever replacing three pointer altimeters or installing them where they had not been in place previously. They didn't seem to have the funds available to modify the whole U.S.A.F. fleet however, which was probably why Captain Milauckas's letter didn't also enclose a cheque!

Flying around Scotland and also on the trunk routes to England often meant coming into contact with all kinds of unfriendly weather.

The severity of the weather at times is ably illustrated by an incident that took place at Ardeer on the Ayrshire coast where I.C.I. once had a large concrete cooling tower. This gave a strong permanent blip on the Prestwick radar and, since its exact location was known, it was very useful for calibrating the radar. One dark stormy night the blip suddenly disappeared. The Prestwick control tower phoned I.C.I. to enquire if their cooling tower had blown down. 'I don't think so,' came the reply, 'but I'll go out and have a look.'

After a short lapse an exclamation of 'My God, you're right!' came across the line. The gale had caused a vacuum within the tower and it had imploded into a nice heap of rubble and that was probably why nobody had heard it.

Eric maintains that he is probably one of the few pilots who, with full approval, has executed a steep turn over the Control Tower at Heathrow flying a Viscount full of passengers. He was due to take the last Viscount up to Scotland. Thunderstorms were raging all over the night sky. Knowing that Eric was Flight Manager, the other pilots in the crew room were waiting for him to cancel so it was the last thing he wanted to have to do. As he was starting up the engines, a landing aircraft reported that he had approached from the direction of Reading and there was a nice clear run between storms from that area.

The take-off run was towards London and, as he was opening up his engines for take-off, a flash of lightning hit the far end of the runway. The Tower agreed when he asked for permission to turn as quickly as he could. He was at a safe height when he reached the Control Tower which is half way down the runway, and he did a steep turn over the Tower and fled to the west as quickly as possible. There was a clear run to the west and he managed to avoid thunderstorms without much trouble, reaching Glasgow without further difficulty.

Eric has never been struck by lightning, although his son, Angus, has experienced this frightening sensation. It is unlikely that serious damage to an aircraft can be attributed to lightning but damage can be inflicted by the turbulence that is created by a thunderstorm.

Sitting in a metal aeroplane is like sitting in a 'Faraday's cage'. The electricity stays on the surface. At night sparks are seen flashing across the windscreen, and in propeller aircraft such as the Dakota, the props are outlined by a circle of 'St Elmo's Fire'. Eric's theory is that aircraft are not struck by lightning but that they short circuit the cloud. Passing through a positive part of the cloud, one collects a strong positive charge. If the aircraft then passes into a negative part of the cloud it discharges with a flash and a bang usually leaving a few small burnt holes where it discharges.

Eric recalls being in the tower at Renfrew as a Vanguard was arriving during a thunderstorm. As the aircraft passed overhead before descending, there was a simultaneous flash and thunder clap. 'That hit me!' came the captain's voice over the radio transmitter. On landing, no noticable damage was found, as expected, but the plaintive voice of the skipper lingered in Eric's ears.

Even without the thunder and lightning the weather can give cause for reflection, which is exactly what Eric was doing one day during a Dakota flight from Glasgow to Belfast in heavy rain. Behind the propeller on the Dakota is the radial engine gear-casing on top of which is the constant speed unit. Eric was passively watching rain drip down from the constant speed unit to the gear casing when he woke up to the fact that this was happening behind a revolving propeller pushing his passengers and him through the air at about 180 m.p.h. There must have been a dead spot right behind the inner part of the propeller. A similar strange air flow happened with another aircraft, the Scottish Aviation Twin Pioneer.

Scottish Aviation were endeavouring to sell the Twin Pioneer to B.E.A., expounding its short take-off and landing capabilities as making it ideally suited to the Scottish island services. Eric arranged a trial with their Test Pilot, Captain Capper, and he proceeded to Prestwick. They went aloft in the prototype and, when airborne, Eric noticed that the port engine was covered with wool tufts. This was an old trick used by manufacturers when they are not quite sure how the air is flowing over the surface of the aircraft. 'What's happening?' enquired Eric. Capper explained that they had been experiencing trouble with overheating in the port engine and when they covered it with wool tufts they discovered that the propeller was acting as an extractor in the engine area and the air was flowing from aft to forward. The cowling and baffles had to be redesigned to prevent this from happening.

During one flight to London-Northolt, a ship's captain asked if he might visit the flight deck. Eric invited him forward but the cloud was down to 600 feet so Eric loaned the sea captain a spare headset while he concentrated on his radar approach. After landing, the ship's captain commented on how little speaking was done. Just the First Officer calling out items from the checklist and the radar operator on the ground saying, 'You are on the glide path. . .you are on the glide path. . .you are ten feet below the glide path. . .you are back on the glide path again. . .' He was most impressed that someone on the ground could tell the crew that they were ten feet too low when they were flying around in thick cloud.

Another ship's master, Commodore Donald Murdo MacLean of the Queen Elizabeth, joined Eric on the flight deck of a Viscount 700 in October 1961 while en route from Stornoway to Renfrew. He was a Lewisman and he had just completed a special journey home to be presented with the Freedom of Stornoway, one of a very small and elite band of men to be honoured by the town in this way. Commodore MacLean had joined Cunard as a cadet after leaving the Nicolson Institute in Stornoway at the age of seventeen. Naval service during the Second World War had seen him on the bridge of a captured German vessel, the 'Bianca', for over a week as he brought it to port. On another occasion he was torpedoed off Rockall and clung to wreckage through the night until rescued. For sinking a German U-boat he was awarded the D.S.C. He had reached the pinnacle of his profession by now commanding the world's most famous liner.

A tall man with a twinkle to his eye, he pointed to a twitching dial marked 'TRIM' on the pedestal of switches, knobs and levers located between the two pilots. 'What's that?' he asked.

Eric explained that the automatic pilot is designed to snap out if something goes wrong or it is overloaded such as could happen if all the passengers moved to the rear of the aircraft. In this situation the auto pilot would be working overtime trying to stop the aircraft doing a steep climb. If it snapped out, the aircraft would do a violent jerk upwards. This, while being harmless, because one could take over quickly, would be somewhat upsetting to the passengers. The trim needle tells one the actual trim of the aircraft and by use of the trimming wheel one can keep it in the middle and the auto pilot can become disengaged without any noticeable effect.

'So you have that trouble too,' was his reply, to Eric's surprise. His small sixty seater Viscount having something in common with the Queen Elizabeth was something Eric had never expected. It transpired that she had been sailing quietly across the Atlantic on automatic pilot, the ship gently rolling. A bucket or some similar object broke loose in the engine room. On its way across the engine room floor it knocked a lot of electrical switches and the ship did a full helm turn to one side. She heeled over to an alarming degree with several thousand pounds worth of damage and with several people being hurt.

Towards the end of Eric's flying career, the Met Office started to produce Significant Weather Charts. The purpose of these was to indicate areas where bad weather would be encountered. Eric maintained that they had got it wrong as he was not interested in the bad weather but wanted to know where the good weather lay. If the weather was impossible at the destination the trip would be cancelled anyway, but if it was poor the important factor was the availability of an alternative airport. The philosophy was to carry adequate fuel to reach your destination, plus enough to reach a safe alternative, plus forty-five minutes for the wife and children.

When Eric knew that the weather was perfect and an alternative destination was not going to be required, a bit of mischief would slip in. Flying from Glasgow to London, he would list Prestwick as his alternate! When a harrassed traffic boy would come and ask him to check his choice of alternate, on account of a full load of passengers and a hold full of cargo, Eric would big-heartedly substitute Manchester.

It has been said that a pilot will be more careful when he has a lot of passengers with him than if he is on his own. Eric believes the opposite is the case since the demands of passengers with appointments to keep will often have a pilot take to the air so long as the weather is within the minima, but in conditions that give him no pleasure from the flight at all. One passenger, who had been stranded at Wick for a couple of days because the east coast haar had engulfed the area, wrote to B.E.A. bitterly complaining and suggesting that the captain would have been in the back office having a good old laugh at the frustration of his passengers. The letter landed on Eric's desk for a reply and he pointed out that if the captain was keeping out of the way it was because he had to make his decision on whether to fly on the basis of the prevailing conditions and not on emotion. Once a pilot started to take an aircraft up in bad weather because he felt sorry for his passengers he was liable to take unnecessary risks.

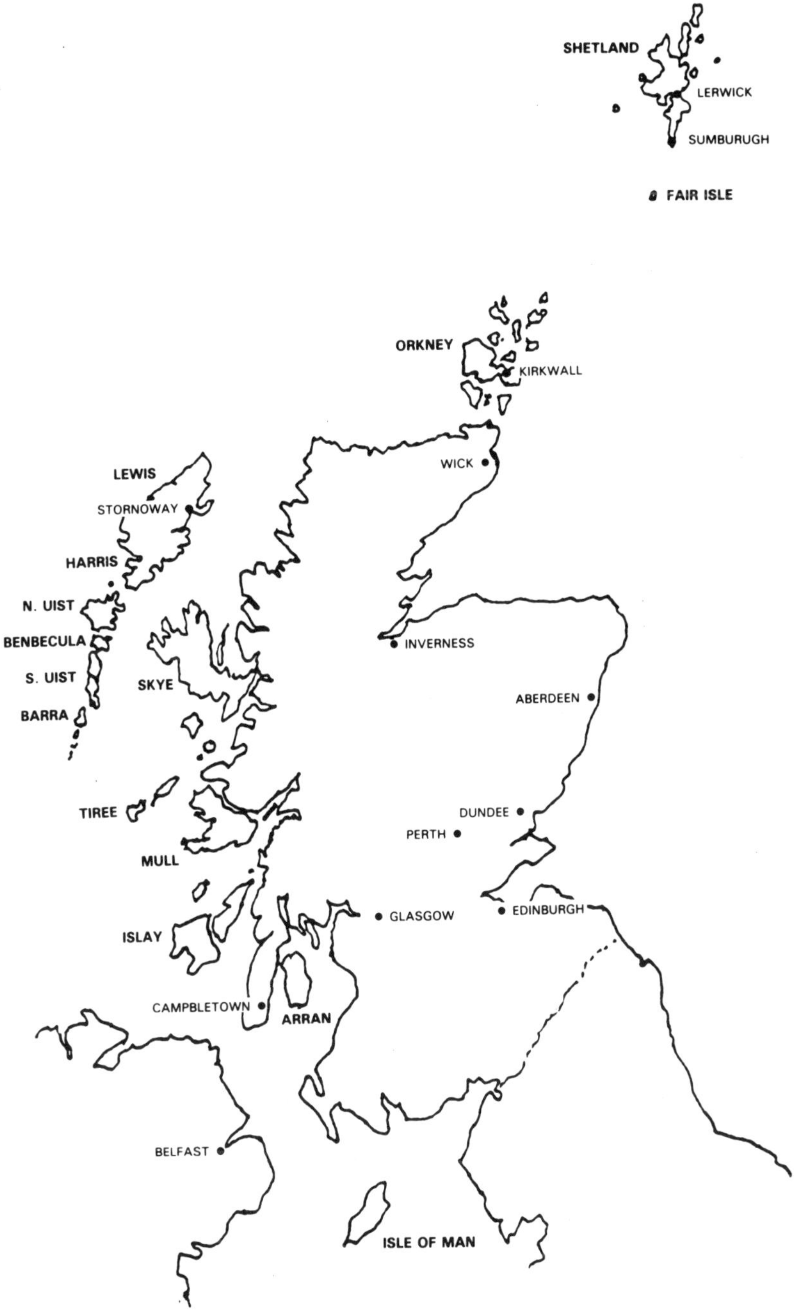
SHETLAND
LERWICK
SUMBURUGH
FAIR ISLE
ORKNEY
KIRKWALL
WICK
LEWIS
STORNOWAY
HARRIS
N. UIST
BENBECULA
S. UIST
BARRA
SKYE
INVERNESS
ABERDEEN
TIREE
DUNDEE
PERTH
MULL
ISLAY
GLASGOW
EDINBURGH
CAMPBLETOWN
ARRAN
BELFAST
ISLE OF MAN

# In Foreign Parts

Eric having survived attacks by bow and arrow across the North Sea in 1947 during the Vickers Viking proving flight, had both this aircraft and the Pionair (Dakota) as the mainstays of the Scotland-London trunk routes for the next few years. Renfrew operated the first southbound flight and the last northbound flight each day so it was arranged that the Scottish pilots also did some Northolt-Le Touquet flights to give them some continental experience (and let them purchase some duty-free goods). However by 1953 the four-engine turbo-prop Vickers Viscount had ousted these aircraft from the London routes. Indeed the Viscount had temporarily appeared on the Edinburgh-Northolt service as early as August 1950 when it had carried 1,815 passengers during that month, many of them being visitors to the annual Edinburgh International Festival.

The Viscounts were flown by London-based crews leaving Renfrew with its Dakotas on regional services. The Vikings had now been phased out. It was not an ideal situation for any aspiring pilots seeking to broaden their experience and Eric arranged to satisfy this demand by having a trip to Paris (Le Bourget) incorporated into the Renfrew schedules to Birmingham and Manchester.

By the time the flight would reach Le Bourget the duty day was already too far advanced for the same crew to fly the return service. The crew would therefore make a highly acceptable night-stop in Paris and then operate the next day's inbound aircraft. A typical two days work would follow a pattern like this one, extracted from Eric's log.

| | | | | | | |
|---|---|---|---|---|---|---|
| 18 May 1953 | Pionair G-AMKE | Renfrew | dep 0612 | Manchester | arr 0738 | Flying time 1hr 26min |
| | | Manchester | dep 0911 | Birmingham | arr 0945 | Flying time 34min |
| | | Birmingham | dep 1010 | Le Bourget | arr 1222 | Flying time 2hr 12min |
| 19 May 1953 | Pionair G-AJHZ | Le Bourget | dep 1333 | Birmingham | arr 1552 | Flying time 2hr 19min |
| | | Birmingham | dep 1611 | Manchester | arr 1654 | Flying time 43min |
| | | Manchester | dep 1951 | Renfrew | arr 2130 | Flying time 1hr 39min |

To say that the crews were spoilt by this arrangement would be understating the situation. Eric, living down the coast in Largs, found that he was spending more time in Paris than in Glasgow to the extent that he had to think carefully before crossing the road when he did make a rare venture on to the busy streets of the Scottish city. A colleague experienced similar problems and remarked that, on bumping into another pedestrian in Glasgow, he had said 'excusez-moi' before realising that it was the wrong response for these surroundings.

One of the London-based pilots had been regularly night-stopping in Paris until the schedules were changed and his lay-overs were switched to Spain as a result. In the restaurant one evening he demanded, 'Café au lait, s'il vous plaît.'

The Spanish waiter gave him a puzzled look but didn't dispute the request. Soon he returned, laid down a cup of coffee, then with a click of his heels and the flourish of an imaginary cape, shouted 'Olé!'

Pilots returning from holiday abroad, wearing ordinary clothes, got one of two reactions from Customs when they spotted that they were dealing with flying staff. One was 'You'll know the score,' followed by a nod through. The other was a thorough going over. Customs at Birmingham were fairly lenient except for one officer who applied the rules very vigorously. During one flight a lady passenger enquired what type of reception she might expect of Customs on arrival at Birmingham. She was advised by the crew that they were quite reasonable except for one fellow. 'We call him Horseface. If you get him, you've had it.'

Later, as the crew followed the passengers through Customs, there was Horseface with his arms buried up to his elbows in this woman passenger's suitcase. She stood patiently while he did this since she had nothing dutiable hidden away. When he had finished she said in a loud voice, 'You must be the one they call Horseface!' He stormed off with a red face while the passenger made her exit, presumably secure in the knowledge that she would have no need to travel through Birmingham again for a long time to come.

One of the pilots intimated to Customs that he had six bottles of Brandy to declare. 'Let's have a look,' said the Customs Officer, a bit taken aback. From one pocket after another a miniature of Brandy was produced until five baby bottles were lined up on the Customs table. The pilot rummaged unsuccessfully for the sixth bottle until his wireless officer suggested that he had noticed him slip it into his case. He gestured to open his case but Customs had had enough and ushered him on out. The sixth bottle was a litre size! No taking bets on how many times that ruse would work today.

It was at Le Bourget that Eric frequently came into contact with American civil pilots. However it was not the first instance as B.E.A. regularly operated through Prestwick during their early days. Shortly after the War morse code had been superceded by radio telephony and it was internationally agreed that English should be the universal language over the R/T. Cloud cover in weather reports was always given in eighths. For example, if the sky was half covered with cloud, it was reported as four eighths. It was soon discovered that some nationalities had difficulty pronouncing the 'th' in eighths so it was agreed to change from eighths to octas. An American aircraft was coming in to Prestwick and the Captain had obviously not kept himself up to date. After asking for a repeat of the weather, he called, 'Say, what's an octa?' A very English voice from another aircraft came over the air with, 'An octa is an octapus with the pus taken out of it!'.

On the overnight trips to Paris the crews stayed at a hotel in the city. One morning the crew bus had come to pick them up when Eric was nobbled by two American ferry pilots who wished to hitch a lift with them to Le Bourget. Conversation developed and since the B.E.A. crew were flying Dakotas, it reminded one of the Americans of a Dakota flight he had done to Bluei West in Greenland. The airfield is at the head of a fjord and the approach is along this arm of water with precipitous cliffs rising on either side. Once the approach has been started, pilots are committed. He recalled one particularly hairy approach but said, 'It was alright for me. I was only the co-pilot. My wife was at the controls!' — He had married a British girl ferry pilot.

During his time as an amateur pilot, one of the Americans related how he had taken up the offer to join a friend who was a commercial pilot 'up front' during a trip on one his flights. It turned out to be a foul day but he didn't wish to pass up the opportunity of witnessing how the professionals coped with this kind of weather since the aircraft was bucking all over the place. He fought his way to the front of the Dakota, bracing himself off the roof as he advanced forward, while thunder and lightning rocked the plane mercilessly. When he reached the cockpit he saw his friend with his co-pilot, their heads together and serious expressions on their faces. Upon getting closer he discovered that it was not the weather that was causing the worried looks but the problems of an absent colleague — 'I told him you could never trust a red head,' the co-pilot was confiding!

Eric commenced his own conversion to the Vickers Viscount on 13th October 1956. Two weeks ground course was followed by a week in the Viscount simulator. Only then was he permitted to fly an actual Viscount with two weeks of 'circuits and bumps' at Stansted. The circuit took him over his old school, Bishop's Stortford College. His son, Angus, was studying there at the time but the concentrated schedule of day and night flying didn't allow the opportunity for a visit. This week was followed by line training flights to Palma and Milan. Next came a six day trip which went London-Rome-Athens-Nicosia-Beirut-Nicosia-Athens-Istanbul-Athens-Rome-London. A few days back at Renfrew to fly a Dakota to Inverness, Stornoway and Benbecula and then Eric was off to Lisbon following this with another multi-sector trip to Beirut. These flights were all in the capacity of First Officer but a trip from London to Zurich on 12th December was culminated by Eric taking command to bring Viscount G-AMOA on up to Renfrew.

Several Viscount trips to Zurich followed during the next few weeks being dove-tailed into Renfrew-London services. On one of these the actor John Mills boarded the aircraft with his family. From the cockpit Eric observed him carrying a small girl, her leg in plaster, whom he presumed to be Hayley after a mishap on the ski slopes.

The Viscount was soon employed to operate the Renfrew-Manchester-Birmingham-Paris service. However on 5th December 1957 the service had to be aborted when fog closed Manchester, Birmingham and Paris Le Bourget. Eric diverted to Liverpool where he waited for four hours then cancelled his flight since it was becoming obvious that Le Bourget would never clear. During this time Manchester cleared and numerous other aircraft were diverting there and disgorging large numbers of stranded passengers while airfields further south remained closed. Eric tried to phone Manchester to see if he should go there as he guessed that they would have a shortage of pilots who were not grounded because of having completed their legal duty day. However the disruptions had caused the telephone lines to become overloaded and seize so Eric decided to fly from Liverpool to Manchester on spec. His arrival was given a warm welcome and his Viscount was promptly despatched back to Renfrew with a full load of Glasgow-bound passengers who had already resigned themselves to being stranded at Ringway for the night.

On another Paris flight, this time from London, Eric had George Rayer, whom he had surplanted sometime earlier as Flight Manager at Renfrew, on board. George was now a Training Captain at London and on this occasion he was supervising Eric in the capacity of First Officer. 'We reached our cruising altitude as we were crossing the coast but the Viscount wouldn't gather speed when we levelled off. George, after a while, said, 'I think we would go faster if you raised the flaps.' We hadn't raised our take-off flaps. I countered with, 'That's a First Officer's duty!' The matter was dropped.'

On 7th July 1959 Eric flew Viscount 802 G-AOHM from London to Milan with Princess Alexandra of Kent on board. It was a hot day and in these conditions it was not possible to gain a high altitude in sufficient time to go directly over the Alps on the return from Milan. A lower altitude was therefore selected but this required a refuelling stop at Zurich.

During a Viscount flight the following year Eric had a door seal start leaking at 11,000 feet while en route from Belfast to London. The cabin air escaping caused a high pitched noise to howl through the aircraft until it was brought down to a lower height to depressurise. This was quite a common fault with the early Viscount 700 series. Cabin staff became quite adept at stopping the escaping air using wet paper towels which were sucked into the hole. But this had no bearing on the choice of aircraft for Eric's next Royal flight.

This took place on 24th May 1962 when Eric flew Princess Alexandra from London to Stockholm in Herald G-APWC — the call sign 'Whisky Charlie' being an unfortunate choice. Unlike other occasions when members of the Royal family had been passengers on Eric's flights, this occasion was an Official Royal Visit. B.E.A. Pass No. 1 was benevolently granted to Eric as Captain which stated 'The Bearer is permitted to approach or board the B.E.A. Herald aircraft carrying the Royal Party.' The Herald's route took it over Amsterdam, Groningen, Copenhagen and Malmo. The visit lasted for almost a week with the return leg from Stockholm taking place on 29th May.

Eric had a sister, Margaret (but always known as Margot), six and a half years older than himself and a brother, James (but always called Jim), who was four years his elder. Jim had suffered brain damage as a baby. He had his own house in Wokingham and worked for many years as a cleaner at Arbrafield Barracks but he required looking after. Money had been left in trust by Eric's father and a family would share the house rent-free in exchange for looking after Jim. Eric paid regular visits to see that this was being done properly and to ensure that he had anything that he needed. Jim once fell and broke his thigh resulting in a spell of hospitalisation.

In the crew room at London Airport, Eric gave friends money to send postcards to Jim from all over Europe purporting to be from famous women. For instance, he received one, 'How I miss you. When are we having another night together? Sophia Loren.' The thoughts of the nurses at the hospital are not recorded but Jim got a lot of fun and prestige. He was not a simple person but suffered much frustration as a result of his handicap. Eric has always been aware of his own good fortune and this was brought poignantly home when Jim confided, 'I wish I was like you.'

Eric's early Viscount trips to Cyprus and Beirut must have made an impression on him for later he was to return on holiday with Eleanor to the former on several occasions. Her one and only visit to Beirut, however, was very short-lived. 'Beirut was at that time a wonderful city. We were having a holiday in Cyprus when B.E.A.'s European Division had a strike. I sent a signal to Glasgow to let them know that my return would be delayed. At seven o'clock the next morning, while my wife and I were still asleep, Olympic Airways rang up saying, 'B.E.A. must have you back as they want you to stand in for a sick pilot. Can you reach the airport by nine o'clock as we have arranged for you to have a seat on the flight deck.' We were staying in a hotel at Kyrenia on the north coast. 'What about my wife?' I asked. When they said there was no provision for her I told them to forget it.'

After breakfast, we went up to the B.E.A. office in Nicosia where we were told that they could get us to Beirut on Cyprus Airways and on from there by B.O.A.C. I readily agreed because I knew that the Cyprus Airways service usually missed the connection and Eleanor and I would have a pleasant day in Beirut at the firm's expense. Off we set the next morning and sure enough missed the B.O.A.C. connection. We had hardly entered the terminal building when my name was called on the public address system. 'Don't waste any time. We have persuaded Iranair to hold back their aircraft for you.' So my wife's only visit to Beirut was a quick dash across the tarmac followed by porters racing after us with our luggage.'

During another holiday, to Israel, Eric and Eleanor visited the border with Lebanon. When he pointed his camera at a tank stationed at the border, a grim and grossly overweight Israeli soldier lurched over from a tank making what Eric presumed was a grab for his camera. Instead he slipped off the lens cover which Eric had forgotten to remove and then broke into a broad smile. Eric's obvious alarm must have made the soldier's day! It was not this incident but their visits to the Wailing Wall and to Caesarea that were to mar their visit. 'Eleanor and I were saddened when, before we were allowed near the Wall, her handbag had to be searched. Nothing seems to be sacred these days. When we visited the ancient city of Caesarea, a party of schoolchildren were being shown around. The schoolmasters who were escorting them were all armed with sub-machine guns.'

One of Eric's First Officers left B.E.A. to become a Captain with a charter airline based in London. As a sideline he and his wife ran a laundrette. One day when he was home, he was doing the shopping for his wife and went into the local grocers. The shopkeeper asked him whether he had any interesting trips lately and he replied that he was just back from the U.S.A. where he had been to both New York and Washington. 'Oh, how I envy you. Here I am stuck behind a counter while you travel the world. Have you any interesting trips coming up?' 'Yes, next week I am off to Singapore and there is a good chance that I may have to go on to Australia.' 'Oh, how I envy you,' repeated the shopkeeper.

Having collected his shopping and left the shop, he suddenly remembered an item that he had omitted. He arrived back just in time to hear one of the customers say, 'Who was that chap?', and to hear the shopkeeper reply, 'He and his wife run the laundrette down the road. He is a headcase and thinks he is an airline pilot.'

Some twenty odd years after being a guest of D.N.L. on their Junkers Ju-52 floatplanes in Norway, Eric was in Finland as guest of Finnair being shown their operations to beyond the Arctic Circle. The 'Circle' consisted of a post each side of the road marking the imaginary line. There was also a log hut selling souvenirs, some under-nourished looking reindeer and a post office from which you could mail cards franked as posted from the Arctic. Eric was told that Eleanor Roosevelt had visited the Circle but the house in which she stayed was 'down the road a bit'. The Arctic Circle was therefore moved to the house for the duration of her stay!

Eric was hosted by Finnair's Chief Training Captain and in his company they passed through a forest while going to board the bus to the airfield. They observed some lumberjacks working in the forest and one of them, an enormous man whose hands seemed to be a foot across, waved to them as they passed by. The Training Captain laughed and said that this mountain of a man reminded him of one their captains who was of similar stature and could only squeeze into the cockpit of his aircraft with difficulty. On one occasion a stewardess had been having difficulty with a drunk passenger who refused to strap himself into his seat. She called on this captain for assistance and he came aft, the sheer bulk of his massive frame plunging the cabin into darkness. He tapped the rogue passenger on the shoulder and said, 'Don't give us any trouble or I shall have to get my co-pilot to you. Now he's a big fellow!'

When not flying regularly, as was the case with Eric in his role as Flight Manager with B.E.A., it was difficult to gain enough flying practice on the Viscounts, now the mainstay of the fleet, to sail through the mandatory six-monthly checks. To get around this problem Eric would periodically go to Berlin for a fortnight where he would fly the corridors of the German internals without the distractions of office work and telephone calls. Berlin was still officially occupied by the Allies and German aircraft were not permitted to fly down the corridors into Berlin. B.E.A. flew these routes for the U.K. and France while Pan Am operated on behalf of the Americans. B.E.A. never based any crews in Berlin but sent them there on stints of two weeks at a time.

This escape to Germany from the telephone wasn't totally foolproof. One evening Eric returned to his hotel to be told that there was a telephone call from Scotland. He immediately thought of the calamities that may possibly have occurred at home. However it turned out to be the B.B.C. wishing to enlist his support for a programme that they were doing on flying in Scotland. They received a reprimand for the scare that they had given him.

The hostesses flying the B.E.A. German internal routes were naturally German. Eric recalls one of them for the manner in which she informed the flight deck that all the doors were secured and passengers safely strapped into their seats. With a Marlene Dietrich voice, Eric frequently commented that she was the only person who could impart this mundane information while making it sound like a seductive love call.

The flights to Berlin used Templehof which was being run by the Americans. It was therefore American operators who gave the radar talk-down. They had devised a procedure for use if an aircraft's compass had malfunctioned by instructing the aircraft to turn so many degrees to the left or right.

Coming into Tempelhof one clear night, they asked Eric if they could do a practice run, using this procedure, with one of their trainees. 'The difference between East Berlin and West Berlin was very noticeable. The West was a blaze of light which was abruptly cut off when one passed over the Berlin Wall into the East. We normally came over East Berlin when approaching to the west. Well, radar headed me East and then seemed to forget me. The countryside got darker and darker and I was just beginning to think I'd better take charge and head back when I heard in the background, 'Tell him to turn...for goodness sake, tell him to turn.' I eventually was directed back to the airfield but at right angles to the landing direction. From there I took over and I must say that, after I had landed, the Tower rang me up and apologised.'

Pan Am flew into Berlin for the Americans and, as there were many parallel services, there was a certain amount of rivalry as to who touched-down first. 'We were somewhat hamstrung by a company rule that we must always approach I.F.R., that is Instrument Flight Rules. Pan Am were not so inhibited and on a clear day would go Visual Flight Rules and cut all the corners.'

'On one such day I was coming in I.F.R. when, on final approach, my blind landing instruments went off line. I landed and on my landing run started fiddling with them to try and get them back on line. As a result I was zig-zagging a bit down the runway and nearly stopped. Just then the Tower said, 'Will you please clear the runway without delay.' As he said it a Pan Am aircraft roared over my head, overshooting because I was still on the runway. It transpired that he was sure that I was deliberately baulking him.'

After the Pan Am flight had unloaded, an enormous angry American stalked across to me. My First Officer did a quick vanishing act anticipating a punch-up on the tarmac. I explained to this angry Captain that I was unaware he was coming in behind me. His face cleared and he said, 'Of course. You were on a different frequency. . . shake!' So we shook hands and all was well. I should explain that when a pilot cancels his I.F.R. plan and goes V.F.R. he is changed to another radio frequency.'

# Running an Airline

Eric became B.E.A.'s Flight Captain at Renfrew in July 1947. When he moved over to become a full-time air ambulance pilot in 1968 he was Flight Manager-Scotland. A few variations of his title had appeared during the intervening years but what they all effectively meant was that he was the airline's Chief Pilot in Scotland and as such he was responsible for running a flight of about thirty crews, or sixty pilots.

On 1st July 1953 Eric was appointed as a Senior Captain 1st Class. The title didn't mean a lot, according to Eric, but as postings were done on a seniority basis it could have proved useful if he had ever wished to move to another station. On 1st October of the same year he was made Master Pilot No. 166 by the Guild of Air Pilots and Air Navigators (G.A.P.A.N.).

Eric also holds the Livery of the Guild of Air Pilots and Navigators. From this it follows that he is a Freeman of the City of London since that is a requirement in order to become a Liveryman in a London City Guild. It is always emphasised that being made a Freeman of the City of London is an entitlement, not an honour. The honour is being appointed to the Livery of a City Guild. (It is interesting to note that a Sir Samuel Starling, no relation to Eric, was Lord Mayor of London in 1669. Three years earlier it had been recorded that Sir Samuel had rewarded thirty able-bodied men for helping remove his personal effects out of the reach of the Great Fire of London. For their efforts he gave them two shillings and sixpence — to be shared among them!)

Eric has also been a Fellow of the Royal Meteorological Society since the early 1930s because, he maintains, the Society were happy to make a Fellow of any pilot who held the then 'B' Pilot's Commercial Licence. Eric claims to be 'somewhat ignorant' about meteorological matters having spent his working life coping with weather rather than trying to understand its complexities. However he has learned his own weather lore over the years. For instance, when Thunderstorms (Cu. Nims.) were around, he would ask to see the Tephigram prior to flying. Cu. Nim. clouds are enemies of pilots because of the associated severe turbulence. The Tephigram is a complicated graph prepared by the Met. men which Eric swears he only vaguely understands but he soon became adept at picking out vital information about these fierce storm clouds, in particular the height range and intensity.

For example, one day he noticed that, according to the Tephigram, the Cu. Nims. around were unlikely to produce much turbulence. Flying up The Minch, a large well-formed thunder cloud was on his path. His co-pilot became unsettled when he realised that Eric was going to fly through it rather than around it. 'Don't worry,' said Eric, 'It's harmless', and harmless it was, with little or no turbulence. The co-pilot was impressed that Eric could predict this with such conviction. Eric told him how it was done.

By June 1952 Eric had logged up 8,000 flying hours. He reached 10,000 hours in July

1960. While his hours were not inconsiderable, a number of his colleagues, if they had been around as long as he, would have notched up higher tallies. Pilots operating over long distances automatically had a big advantage over those who were operating short sectors. But in his role of Flight Captain at B.E.A. Eric had to spend a considerable amount of time behind a desk. This situation had in fact started as early as 1942 when Eric became a 'part-time pilot' with the award of his first command, Flight Commander of 221 Squadron. All his subsequent positions involved him in an adminstrative role that curtailed the time available for flying. Indeed getting a reasonable amount of flying, an integral part of the job at B.E.A., could present problems.

'On more than one occasion I would be making my way to the aircraft with my secretary running after me with a sheaf of letters requiring signature. It was only after we had been airborne for a few minutes that the office was truly left behind and flying could take over. Even then I was never totally immune. One day while I was crossing the Solway Firth en route to London, the tower at Prestwick called to say that they had a message from B.E.A. 'We want one of your captains to take a flight to Kirkwall but his route experience has run out by a week. Will you authorise him to go?' There was just no closing the office door!'

Eric found that running his flight of sixty pilots presented two problems. One was how to control staff and impress one's personality on them when they spend their working day miles away. They other was how to ensure that they were maintaining the high standards demanded by a responsible airline.

The first problem is difficult because pilots only arrive at the airfield with sufficient time to prepare for the flight and when they return they want to get home and not hang about chatting up the boss. That only leaves the standby crews available for closer scrutiny. With the Flight Manager's own commitments to flying and meetings, he is only available in his office for up to three days a week. That means that with the best will in the world he can only expect to meet every member of his flight once a fortnight.

Ensuring that standards and laid down procedures were being adhered to was more easily accomplished through the six monthly checks performed in the presence of a training captain. The only way of doing well during the checks was by operating correctly at all times so that it became a habit. Eric too underwent these checks and his experience of a passenger dying on him during an overshoot at Benbecula under test conditions is recalled in an earlier chapter. In that instance the passenger was pronounced dead on arrival at Stornoway so there were no problems in deciding his place of death for the death certificate. Another crew flying a Dakota from Palma had to resort to more desperate measures to avoid bureaucracy when a man died of a heart attack before they had reached the Spanish mainland. At Barcelona he was tied up in a parcel and then flown back to Glasgow in the luggage hold as freight so as to bypass long delays while the Spanish police and legal system became involved in what would surely have been a leisurely fashion.

The first item in the aircraft check-list is 'First Officer Briefing'. If Eric was flying with an unfamiliar First Officer, perhaps one from London doing a locum, he always emphasised in his briefing that though he was a very senior captain with years of experience, he was quite capable of occasionally doing something stupid. It was the First Officer's job to cross check everything and not be afraid to point out anything that Eric was doing which he considered incorrect — tactfully but firmly. 'Excuse me, Sir, didn't he say turn right, not left?' There have been quite a few crashes because the First Officer was frightened to open his mouth. The beauty of having made this clear was that if Eric did do something wrongly the First Officer didn't know whether this was a real error or Eric just putting his attentiveness on trial.

One of the less savoury aspects of Eric's job was dealing with unsuitable pilots. This was eased by an arrangement with B.A.L.P.A., the pilots' union, whereby new pilots would be on probation for their first six months and could be dismissed without comeback. One candidate proved to be totally unsuitable. Before taking action, Eric also discovered that he had entered the Royal Australian Air Force as a pilot but had quickly been dismissed as unsuited. Despite this, he had gone to a flying school and gained a Commercial Pilots Licence, presumably sponsored by his wealthy father.

His father turned out to be a friend of Lord Douglas, Chairman of B.E.A. at that time, and he demanded of him on what grounds the Flight Manager in Scotland could turn down his fully qualified son. 'This went down the line and my boss, Paddy Louden, the General Flight Manager, came to see me for an explanation. Now Paddy had four sons so I said to him, 'If one of your sons had been turned down as unsuitable for aircrew by the Australian Air Force and unsuitable as flying staff for B.E.A., wouldn't you be keen to get him out of flying and into something he could be good at, as soon as possible?' I said that I had explained to this boy that he would be unlikely to become anything more than a second-rate First Officer with no hope of ever becoming a Captain. I told him to get out and into something he could do well. I told him that my experience had been the same. A far-seeing boss chucked me out of advertising and helped me get into flying which he knew was my heart's desire and at which I became a success.

'There is more to airline flying than just holding the necessary licence in the same way that different qualities are required of a test pilot or a flying instructor or a helicopter pilot. The Malays in Brunei made very good helicopter pilots. I know because I did a lot of flying with them but I was told that they were poor fixed wing pilots. Why, I never understood.'

Fathers pulling rank was not something which cut ice with Eric. 'Bob Preston, Flight Manager-Manchester, and I had an arrangement whereby we looked after each others Flights during leave. I was paying one of my visits to Manchester when a father came to see me. He said that he had a son who was very keen to become an airline pilot but he hadn't enough 'A' levels to get into the College of Air Training at Hamble where, at that time, all future airline pilots were trained. He said that it was a pity that someone so keen should be passed over and could I pull any strings to get him accepted. In reply I told him about one of our pilots in Scotland. He was the younger son of a farmer who had left school without qualifications to work on his father's farm. As a younger son, he knew he would never get the farm and looked like being a farm labourer all his life. The idea of becoming an airline pilot fascinated him but he didn't have the necessary qualifications. So he went and saw his old headmaster and asked if he could come back to study for his exams. The headmaster agreed, he got the necessary qualifications, went through the College of Air Training and is now one of our First Officers. 'That,' I told the father, 'is what I call motivation. Not someone who waits for his father to pull strings rather than take his future in his own hands. Tell that to your son!'

Eric also had his own philosophy about how to advance one's career. This lesson had been learned from Larkins, the chief space-buyer at his very first employer, Saward Baker. Larkins knew that no one could match his skills and that he was suffering as a result. Unless there was someone who could move into his position with the firm his own advancement was going to be thwarted. Eric's belief has therefore been that, to get on, you must be able to do the job of your immediate superior, and also have someone who can do your own job.

Eric's role as Flight Manager put him in a slightly incongruous position in that he was a member of the management team while also being a pilot. When it came to management/staff negotiations Eric however found that he was firmly management while having to exercise a degree of understanding of the pilot's view. Fortunately he was an ideal person to deal with matters from a fair and balanced viewpoint; although some of the pilots heavily involved with union activities naturally took a different viewpoint of certain situations that arose.

'When Scottish Airways was nationalised into B.E.A. by the Labour Government, one of the stipulations was that a Pilots Panel should be formed at each base so that pilots and management could consult with each other. The pilots appointed a chairman who was immediately dubbed the 'Shop Steward' and the problems consistent with a union type consultation quickly followed. A pilot having been elected 'Shop Steward' felt he was letting the side down if he didn't have something to moan about at the next meeting. In other words, if you are happy in your work you are being exploited. If one looks over the minutes of Pilot Panel meetings, the same niggles come up again and again; meal allowances, car allowances, hotel accommodation and so on. We ran a joint Pilot and Radio Officer Panel.

'At the time I was put in charge, the Radio Officers were on their way out because there was no point in carrying Radio Officers when all air to ground communication had changed over to radio telephony. Just before I was appointed, an urgent ambulance call came up and a Radio Officer couldn't be found. I said I would take it as I had done more Rapide flying without a Radio Officer than with one. The Radio Officers were furious as no agreement had been reached with them about the reduced crew.

'By the time of the next Panel meeting, I had been put in charge. When the subject was raised I reduced the Panel to laughter by very solemnly announcing that I had spoken to the pilot concerned and he had given me his assurance that he would not fly without a Radio Officer until an agreement had been reached. I would add that B.E.A. treated the Radio Officers very well and offered a Pilot's course to any of them who could pass the Pilot's medical.

'There was a certain Pilot in London who would always put the rule book before consideration for the passengers. During the Viscount days, this Pilot found, after a night-stop in Edinburgh, that a gale was blowing down the short runway and it was too strong across the main runway. Operations in Glasgow called me up to see what could be done. One of the boffins down in London had written in the Operational Manual that this runway was unsuitable for Viscounts. He had stated that it was unsuitable because under normal conditions it couldn't be used when carrying a fairly heavy load of passengers. I suggested to this pilot that he therefore take-off empty, fly over to Glasgow and await the passengers who would be taken across by bus. No, he wouldn't do that because the runway was declared unsuitable for Viscounts. 'Not even with the authority of me, a management pilot?' 'No,' he was adamant. 'O.K. I'll come across and do it myself.'

'By the time I reached Edinburgh by taxi, the passengers were on their way to Glasgow where there was another Viscount. So Operations now just wanted me to fly this empty one directly to London. It was a Viscount 800, the powerful one. My dander was up so the moment I had flying speed, I pulled hard back on the stick and with the gale force wind and the emptiness of the aeroplane, I went up at an angle of about 45 degrees. I must have been at about 600 feet when I passed over the end of the runway. The traffic boys just loved it and still laugh about it.

'The work-to-rule Pilot was furious and told his Flight Manager that I should be dealt with. The Flight Manager got the boffins to work out what load, if any, it was safe to carry under the prevailing conditions. The answer was that fourteen passengers could have been carried. That finished that. The London Flight Managers were delighted that I had made this thorn in their flesh look a damned fool.'

Despite being 'management' Eric also had to adopt tactics to ensure that he had satisfactory conditions for the running of the Scottish Flight. If he felt that he had a staff shortfall it was frequently fruitless to solve it by requesting more staff. A less direct approach had to be adopted. Eric found himself in this situation so he submitted to the personnel authorities that he believed that he wasn't running his operation efficiently and requested that they sent some experts to point out areas that could be improved. Eric had of course already gone through a similar exercise himself so he wasn't totally surprised when the experts reported, 'We can't see any areas where efficiency can be improved as you are already doing everything that can be done. You'd better have some additional staff!'

It was occasionally necessary to keep pilots on the right track, especially if they became too friendly with their colleagues in the passenger cabin. One first officer was becoming deeply involved with a hostess. Meanwhile his wife was continually having to seek the company of her girlfriends during her husband's overnight trips away 'on duty'. While Eric was reluctant to intervene in these friendships, this one was becoming the talk of the base and eventually he decided to have a word through the right channels to arrange the transfer of the stewardess to London to cool things down. Eric did not expect any thanks from the amorous first officer for this arrangement but some years later he confided to Eric that his action was greatly appreciated. He confessed, 'I was getting into a situation there that was becoming increasingly difficult to escape. I realised what you had done and owe you a lot of thanks.'

Occasionally a pilot would suffer from the 'twitch', not a condition caused by stewardesses, but a sudden neurotic fear of flying that would sometimes appear out of the blue. In the Forces such a pilot could be quietly grounded but in commercial flying pilots were well paid but often little qualified for anything else. With this in mind the unions would fiercly resist any attempt to ground a pilot unless it could be proved that he was becoming a danger to himself and his passengers.

Eric had a pilot with such a problem which was displayed by his cancelling of services unnecessarily and being short with his co-pilots. Eric could see the symptoms but resisted calls to ground him unless he stopped flying safely. Eventually the pilot came and tendered his resignation. 'Thank heavens for that,' said Eric. 'What do you mean?' responded the pilot taken aback. 'Well, it has been obvious for quite a while now that flying is taking too much out of you,' said Eric. 'Was it really as apparent as all that?' responded the pilot.

Eric had a partial solution to the problem. 'I'm going to tear up your resignation and I am going to send you to the Medical Officer to have you grounded for medical reasons.' Eric then phoned the Medical Officer to give the background to the pilot who was about to arrive in his consulting room. This way the pilot got his sickness pension and £20,000 loss of licence insurance whereas his resignation would have brought him nothing. The pilot subsequently invested this money by putting himself through a management course and so embark on a new career.

Throughout twenty years of running the Flight at Renfrew, Eric enjoyed a good relationship with his crews. However he wasn't totally sure how to react to a newspaper banner headline that appeared on the staff notice board one day. 'Extermination is the only answer to the Starling Problem,' it proclaimed. Someone with a mischievous sense of humour had spotted it above an article concerning the flocks of starlings that used to brighten up Glasgow with song as they homed in on George Square to roost at night but simultaneously didn't do a lot for the appearance of the city's Georgian and Victorian architecture.

Most pilots retired at fifty-five but, because Eric was older than many of his colleagues when the pension scheme came into being, he was not scheduled to retire until he reached sixty. By the time he had reached fifty-five he found that he was beginning to struggle with his flying checks. Being in charge of the Scottish Flight was the major reason for

Fresson's DH-84 Dragon G-ACIT, 'Old IT', which was subsequently spirited away from Inverness by B.E.A. after nationalisation.
*(Courtesy of Peter Clegg)*

The life of the Junkers Ju-52 with B.E.A. was short — although some of that distinctive corrugated sheeting still serves Eric well as flooring in his loft.
*(Adrian Meredith Photography)*

B.E.A.'s first post-war airliner, the Vickers Viking.

The rebuilt DC-3 emerged from the workshops to fly with B.E.A. as the Pionair.
*(RAF Museum/British Airways)*

The unflappable Colin MacPhail, Station Superintendent at Tiree, receives a presentation from Eric to mark his retiral. On the left is Captain David Barclay and on the right is Jimmy Mitchell, Flight Administration Officer for B.E.A. in Scotland.

The Handley Page Herald which took to the Scottish skies on behalf of B.E.A. in 1961. *(Adrian Meredith Photography)*

The Viscount V700, seen here at Dublin in 1952, brought a new era to flying that was appreciated by passengers and crews alike.
*(RAF Museum/British Airways)*

B.E.A.'s fleet serving Renfrew during 1962. Handley Page Herald, Vickers Viscount, Vickers Vanguard, Douglas DC-3 Pionair and DH-114 Heron.
*(RAF Museum/British Airways)*

Eric, as B.E.A.'s Flight Manager-Scotland, in his office at Renfrew Airport.

this since the job rarely permitted him regular flying and he found that his standards were beginning to suffer as a result.

'By the time I was fifty-seven I could no longer make the grade and failed my Viscount check. There were two kinds of Viscount checks. There was the Route Check where a Training Captain flew with you on a trip to observe that you were carrying out all the laid down procedures, and the Base Check where your general handling of the aircraft was scrutinised and you were examined in your handling of engine and system failures. It must be remembered that a Viscount, although slow by modern standards, could travel at any speed from 100 to 300 m.p.h., or five miles a minute. It is said that when you blink your eyes while travelling in Concorde that you go one hundred yards with your eyes shut.

'On a typical check the Training Captain would fail an engine during the take-off run. Having successfully shut down that engine, feathered the propeller, and pulled a very lop-sided aircraft into the air at sufficient speed to have enough rudder to hold against the dead engine, one climbed away straight into low cloud and was on instruments from then on. One had just completed the take-off drills and trimmed out the dud engine when another would fail on the same side. In cloud, with a plane with power half gone, and a plane which was trying to go round in circles while under the strain of a test is not fun. One then had to call up the Tower, arrange to come in on the beam and do a successful beam approach with an aircraft which was very uncomfortable to fly. If the Check Captain was feeling very sadistic, the undercarriage would stick up and the emergency system would have to be used, probably making it necessary to go around for another approach. I would add that this check was always done in a simulator. In the early days, we crashed a Viscount because a Training Captain failed an outboard engine during a genuine take-off run and the Captain under conversion promptly feathered the inboard engine. Even the somewhat forgiving Viscount will not leave the ground with two engines on the same side failed. There is an old saying that if one deliberately stops an engine, another is bound to fail.

'The flight deck of a Viscount simulator is exactly the same as the real thing. It even makes the correct noises and one can hear an engine fail. It is easy to forget that one is flying a simulator. I remember that on one occasion the Check Captain 'threw the book at me'. Everything that could go wrong went wrong. Even the undercarriage stuck up and I had to operate the emergency system to get it down. However I successfully battled my way down against great odds but, as I throttled back to land, the hooter blew. This blows if the undercarriage is not down and locked. There was no hope of going round again with only two engines on one side so, exclaiming, 'My God, we've had it', I waited for the expensive noises. They didn't come — I had forgotten I was flying the simulator. I would add that it was a fault in the simulator this time.'

Eric admits that, after such a long and distinguished career, he took a little time to come to terms that the faithful Viscount was now letting him down. However if Eric might have felt that years were catching up on him at this time, he was soon to be rejuvenated by a return to flying very similar to that of the thirties.

# The Starling and the Heron

## *Aeroplane dash from Islay*
## *Man rushed to Hospital*

*A dash by aeroplane from Renfrew to Islay in order to convey a sick man from the island to the Western Infirmary, Glasgow, was made yesterday. Shortly after 9.30am a telegram was received at the headquarters of the St. Andrews Ambulance Association asking that a plane be sent for an urgent case. The Scottish Flying Club were communicated with, and they in turn got in touch with Midland and Scottish Air Ferries at Renfrew.*

*A plane was immediately despatched and arrived at Bridgend forty minutes after leaving Renfrew. The patient — John McDermid (33), fisherman, Bruichladdich — was got into the machine, which returned to Renfrew, where an ambulance waggon was waiting. He was conveyed to the infirmary, where he underwent an operation for abdominal trouble.*

*The patient, it is learned, is now making satisfactory progress. Normally the patient would have taken the best part of a day to come to Glasgow by the customary boat and motor or train services.*

## *Previous Errand of Mercy*

*While this flight is thought to be the first in Scotland on which a patient has been brought from an outlying part of the country to the city for an urgent operation, it is not the first time that a Renfrew plane has been used on an errand of mercy. Last year an aeroplane was sent off one afternoon from Glasgow to Islay carrying certain medical preparations which were urgently needed for a case on the island. As on this last occasion the journey was made in a very short space of time and without mishap.*

This report in the Glasgow Herald of Monday 15th May 1933, inserted as a small human interest filler, unknowingly announced a historic landmark in the development of Scottish civil aviation. John McDermid was suffering from perforation of the stomach and the threat of peritonitis was high. When Captain J H Orrell landed on the beach near Bridgend in de Havilland Dragon Moth G-ACCZ to whisk the fisherman off to hospital he almost certainly saved McDermid's life. The Scottish Air Ambulance Service had arrived and remote communities were quick to appreciate that access to an aeroplane could represent the difference between life and death in times of illness or accident.

Prior to the formalisation of the air ambulance service with the setting up of the National Health Service in 1947, the aircraft had to be chartered in times of emergency. As the financial burden of such a flight would be too much for one family, communities set up funds for the purpose, running sales of work, concerts and other functions. The resources provided by these funds were soon being put to use around various parts of Scotland and Eric flew his first air ambulance flight while with Aberdeen Airways when he flew a patient from South Ronaldsay to Stromness on 2nd February 1936. This was also the first ambulance flight undertaken by the Dyce-based airline but air ambulance flying was to be a regular interruption to Eric's flying activities from that day on. This included his war-time flying when he flew an Anson from Limavady to Abbotsinch bearing the burned survivor of a flying boat crash.

Much of the air ambulance flying during the fifties and sixties was undertaken by David Barclay, the Northern and Scottish Airways pilot who had been responsible for locating many of the Hebridean airfields and establishing the first air services to many of the communities of those islands. During the Second World War, however, David Barclay had been badly injured when the Dragon he was piloting with a load of senior Army personnel took a nose dive as he prepared to land at Kirkwall.

That he fought his way back to mobility was a major feat of courage in itself, but that he was back in the air within several months of the accident speaks volumes of the type of stuff of which Captain Barclay was made. But an enduring effect of his injuries, a shattered ankle, meant that he had reduced power in his leg. This meant that while he could fly larger aircraft, such as the Dakota, under normal conditions, the damaged ankle presented an insurmountable handicap when the heavier aircraft required to be flown in emergency conditions.

David Barclay's flying career was therefore restricted to scheduled flights on routes such as Tiree and Barra where smaller aircraft were used and to the Air Ambulance Service. It was following an incident involving David Barclay in November 1950 that subsequently saw the Rapide bi-planes, that had blazed the golden age of flight, being gradually retired. Encountering severe icing and strong headwinds en route from Benbecula to Renfrew, the Rapide was slowly being forced lower and lower as it struggled on. After coming dangerously close to the mountain tops over Argyll, the aircraft finally made it home with dwindling fuel supplies having taken over three hours to complete a route that would normally require one hour and forty-five minutes.

Captain Barclay lobbied strongly for a new aircraft to supercede the Rapides and one that was equipped with de-icing equipment. The result was the four engined de Havilland Heron. One great attribute of the Heron was not only that it could fly on three engines but could also take off on three engines. If an engine malfunctioned while the aircraft was on one of the islands it therefore needn't be stranded until a team of engineers arrived with a replacement as Eric found to his advantage on more than one occasion. Also if an engine wouldn't start it was possible to take off on three engines, let the wind assist the fourth engine to start once airborne, and then return to the ground to collect the passengers.

The icing problems with the Rapides meant that night ambulances were only carried out in good weather as it was not possible to creep around the mountains below freezing level in order to reach the islands. However when the National Health Service came into being, it became responsible for medical care in the islands and asked B.E.A. for twenty-four hour air ambulance standby cover. An aircraft was therefore required that had the sophistication of an airliner and the handling qualities of a light aeroplane. The de Havilland Heron with four 200 h.p. Gipsy engines fitted the bill. It was also suitable for operating the daily scheduled service to Barra where the airfield was a sandy bay.

Sand and salt water are not good for retractable undercarriages so Herons with fixed undercarriages were ordered. This caused a bit of fun at Machrihanish, which served Campbeltown and was controlled by the R.A.F. The occasional new controller who didn't know about the undercarriages would give the standard R.A.F. call, 'Clear to land — check wheels down and locked.' To this Eric would reply, 'Roger, wheels down and welded!'

Eric made two modifications to the Herons. The health authorities wanted the aircraft to be used to fly infectious diseases cases in from the islands but the pilots were not happy about this. Eric designed a detachable bulkhead that would separate the cabin from the cockpit. But a ventilation test had to be conducted to establish its effectiveness. The ventilation in most aircraft is from the rear forward so Eric purchased some stink bombs which were let off in the passenger cabin. The smell did not penetrate to the cockpit so the modificiation was passed as effective. The aircraft would be fumigated after such flights before going back into service.

The second modification instituted concerned the fire extinguishers and first aid kit. These essential items are difficult to locate in the event of a crash but in nine crashes out of ten the tail would be unharmed. A compartment was therefore devised in the tail of the Heron into which the first aid kit and fire extinguishers were fitted.

With three thousand ambulance flights to his credit when he retired in 1965 it is easily understood why Captain Barclay was regarded as something of a hero in the islands. When he reached retirement he received numerous honours from island communities who each regarded him as their own. Eric Starling remembers one such accolade which brought with it a little embarrassment for himself.

'When David retired, the people of North Uist and Benbecula contributed towards a presentation for him which had been organised by Dr Macleod. The presentation was held at the local school and my wife and I were also invited. I tried to get out of it because it was David's party and I didn't want to intrude. However they were very insistent that I come. After the usual speeches, David was presented with a cheque for £150. When he had said the appropriate words, a big box was passed down the table and dumped in front of me. It was a present consisting of a very fine Zeiss 35mm slide projector. I was flabbergasted. I had to do some quick thinking and in my thank you speech I recalled that during the War, if a crew put up a good show, even though the Captain may have been asleep in the back it was he who got the medal. As Flight Manager, I said I looked on it as if I were now the Captain getting the medal for all the good work done by the pilots, engineers and nurses responsible for the success of the Air Ambulance Service.'

When Eric relinquished his job as Flight Manager to become a full-time Air Ambulance Pilot he was returning to the type of flying that he had been brought up on and this suited him ideally. Doing his twelve hour ambulance standby shifts at home, he could reach Glasgow Airport comfortably from Largs within one hour of the call out. He always had to be within earshot of the phone so he had a bell fitted to the chimney stack of his house. When the bell sounded it provided some drama for his neighbours while it permitted Eric to work in his garden while he was on call.

The Air Ambulance required to be airborne within one hour of a call. The patient would normally already be under the care of the island doctor, who would place the call for the air ambulance, and there would seldom be the urgency required of a road ambulance dashing to the scene of an accident where the victims may be bleeding to death. An hour was therefore adequate for a nurse to be fetched from the Southern General Hospital while the crew took care of other matters such as filing a flight plan. One night Eric was belting through the deserted streets of Paisley and found that a police car was following him. He decided to carry on without slowing down and the police followed him to the airport, pulling up directly behind him. As Eric got out of his car he asked them not to delay him as there was a sick boy on Tiree waiting for him to fly him in. The police were satisfied and Eric was soon in the air.

In the early days of B.E.A. air ambulance flying continued to be supplementary to the mainstream operation of scheduled services. For example on 1st March 1948 Eric flew a Dakota on the evening service from Renfrew to Nutts Corner, Belfast. Darkness had fallen by the time of the return journey and the weather was closing in on Renfrew. Eric was quite happy to arrive ahead of the weather until met with the news, 'We have had a call from Campbeltown for an ambulance. We have a Rapide ready. Will you take it?' So Eric immediately stepped from one aircraft to another to answer the emergency call.

With the arrival of the National Health Service this *ad hoc* system was changed so that crews were available on permanent standby for air ambulance duty twenty-four hours a day. This was facilitated by the ability of the Heron to operate at night, a task rarely asked of the Rapides which were not well suited to night flying. One crew covered a twelve hour shift from 6.00am until 6.00pm when they were relieved by the second shift which took over until the next morning.

It continued to be at a crew's discretion as to whether they took to the air and this often meant ambulances setting off in conditions that were keeping all other aircraft on the ground. But more stringent rules were laid down after a tragic crash on 28th September 1957 when Heron G-AOFY came down on Islay with the loss of Captain T N Calderwood, Radio Officer Hugh McGinlay and Sister Jean Kennedy. Eric had the awesome task of taking out a second aircraft to collect the patient, who sadly was to die shortly after arrival at Renfrew. In future, ambulance pilots would be required to observe weather minima.

Nurses were effectively cabin staff on the air ambulances and knew their way about the aircraft better than anyone beyond the flight deck. As part of their training Eric, as Flight Manager, would be called in to do a talk and would give them some important tips. One of these was the correct way to provide the Captain with a cup of coffee and this wasn't totally frivilous advice. It was explained that the deep cups were intended to be only half filled to prevent spillage, very important about fifteen minutes out of Renfrew when Goat Fell on the Isle of Arran could provide some bumpy wind currents. The last thing the pilot wanted was an overfilled cup of coffee slopping over him and it wouldn't do the electric circuits of the aircraft a lot of good either if it was spilled over the pedestal.

Ambulance flying provided serious moments due to the very nature of the facility. But there were frequently lighter moments.

Among the serious cases carried was fifteen year old Alex Smith of Breasclete on the Isle of Lewis. Alex had accidently drunk paraquat and he was rushed from Stornoway to Edinburgh on 10th May 1968 to become Europe's first lung transplant patient. On 2nd January 1969 Eric was carrying more poison victims. Four Hogmanay revellers from Campbeltown had drunk fire extinguisher fluid from a lemonade bottle. 'What idiot was responsible for that?' asked Eric. It turned out to have been one of the victims. One of the quartet did not survive the consequences of the prank.

For Eric the most traumatic flight was probably when he carried a twelve year old boy from Barra to Stornoway to have his appendix removed. The young lad, equal in age to Eric's son Angus, boarded the aircraft at Barra in the company of his father. He carried a small suitcase and obviously viewed his visit to the hospital on Lewis as an adventure to be savoured. Three days later Eric flew them back to Barra, the father extremely drunk and open to no kind of consolation, and his small son in a coffin in the back. He had died shortly after the operation.

One ambulance flight which Eric might prefer not to remember took place on 19th May 1964. Dundee had recently built a landing strip on a piece of ground beside the Firth of Tay which had been reclaimed by using it as a rubbish dump. Eric flew there to pick up a woman who had broken her back in a road accident. After refuelling at Edinburgh, Eric flew his patient to Halton which was close to Stoke Mandeville Hospital. Halton was a grass field, slightly downhill, and on this occasion, very wet. It was the first time that Eric had landed the Heron on grass and he had forgotten that wet grass downhill was, unlike a wet runway, very slippery. He concentrated on a doing a smooth landing, braking only after he had gently lowered the nosewheel on to the ground. He therefore started braking rather late and found to his horror that the brakes were having little or no effect on the wet sloping grass. With visions of going through the boundary hedge, he cut all four engines to cancel the small thrust they were creating at the kick-over. As he slowed, the brakes fortunately began to take effect and he stopped just in time. 'I hate to think what would have happened to my poor patient if she had followed a car accident by a crash through a hedge in an aeroplane. Flying is full of traps for the unwary and I was truly ashamed of myself.'

Eric's role as Air Ambulance Pilot took a slight deviation on 15th November 1970 when he proceeded to Aberdeen in Heron G-ANXA. A B.E.A. helicopter had ditched in the North Sea and Eric joined in the search. Using his experience gained in Coastal Command, he scanned the sea with an expanding square search using his Decca Navigator. (The Decca Navigator was very accurate and invaluable for finding Barra in bad visibility — by following the right position line a tiny island could be located at the entrance of North Bay.) Eric spent nearly four hours in the air until the dinghy containing the helicopter crew was eventually picked up by a ship 65 miles SSE of Aberdeen. There was no trace to be found of the helicopter.

Only a small proportion of ambulance flights are extremely urgent, a fact which Eric impressed on the nurses after one incident. He was just coming to final approach at Glasgow when he looked round to see the patient seated on a bedpan. Eric explained for future reference that under such circumstances it was quite acceptable to advise the control tower that they had a small problem on board and could they circle for ten or fifteen minutes. The patient could then be brought down with comfort and dignity.

Not all air ambulance passengers were the patients and their relatives. A young locum serving on Barra thought that the attendance of a doctor on one particular flight might be prudent for the patient's well-being. Sister Anderson, a delightful Jamaican nurse flying for the first time with Eric, then spent most of the journey to Renfrew tending the doctor, who became violently air sick, with even the patient lending a hand to relieve his distress.

Nurses of course could be air sick too. One patient, uplifted from Benbecula, was a retired gentleman of military bearing and he was not going to permit himself to be stretchered to the awaiting aircraft. Poker-backed, he walked out to the aircraft and slumped into a seat. He was an angina case and had to be flown at a low altitude which meant a bumpy flight in the prevailing blustery weather. After they had been airborne for a little time, Eric looked back to see how his passengers were getting on. The scene that confronted him was the patient in his permanent posture of stern dignity sitting with the nurse resting her head on his lap. In one hand he held a strategically positioned air sickness bag while he soothed her by stroking her hair with the other hand.

Attached to the staff at Renfrew was an industrial padre and he asked Eric if he could join him on an ambulance flight. 'Fine,' said Eric, 'but only after you've removed your dog collar. It might not do the patient's morale a lot of good if he sees a priest on board.'

Eric's three years as a full time Ambulance Pilot required him to do twelve hours on and twelve hours off. Such a long period of duty was only legally possible if the standby shifts were done at home. Therefore, after three years of spending so much time at home, flying had become increasingly an interference with his home interests. It proved to be an ideal winding down process in the run-up to retiral so that when he reached sixty Eric was quite glad when he no longer had to stop what he was doing and dash off to the airport.

# Flying to New Horizons

*'I get up in the morning with nothing to do*
*and go to bed at night with only half of it done!'* E.S.

On 6th December 1971 Eric undertook his last day's flying with two trips in the Heron to Islay. The first of these was an ambulance flight to collect two expectant mothers so that they might have their babies delivered in one of the Glasgow hospitals. While on the ground at Islay Eric was telling the women that this may well be his last flight but, throughout forty years of flying, he had never had an inflight birth.

Just at that moment one of the airport staff came rushing from the terminal building, 'Could you call in on Machrihanish on your way back and collect a woman who is in labour?' 'You might get your inflight birth yet,' joked the women, 'Just keep circling around until the baby is born.' However the patient from Campbeltown managed to hang on to her baby until she was safely in hospital in Glasgow and Eric never did get a birth on board one of his aircraft.

As Eric hung up his captain's cap for the last time he had flown 12,548 hours and 5 minutes. During forty years he had flown forty-five different types of aircraft ranging from single engine biplanes to multi-engine airliners.

At the time of Eric's retiral pilots could fly all types of aircraft until the age of sixty but after that age a restriction prevented them from flying the larger types. It had become evident that, after the age of fifty-five, a significant number of pilots were having difficulty in passing either their six monthly flying checks or their six monthly medical checks. In order that the majority of pilots could expect to retire on a full pension, it was agreed with B.A.L.P.A., the pilots union, that their members would pay slightly higher premiums so that they could retire on a full pension at the age of fifty-five.

When the scheme started, Eric was already older than many of his contemporaries so that he would have been well into his sixties by the time he could have contributed long enough to retire on the full pension. However Eric was able to 'buy' some extra years pension so that he would have his full entitlement at the age of fifty-nine. As the time approached he was encouraged to continue flying until he was sixty. 'By the time the Union find out, you'll be sixty anyway,' his boss told him. So, having first flown at nineteen, Eric continued until his sixtieth year by which time he was the most senior pilot in B.E.A.

While he would never say it while still flying, Eric could now think about hanging up his captain's hat and at last be able to boast, without tempting providence, that he had never had a serious accident. 'Throughout my airline flying I have been trying to take the adventure out of air travel. I feel I have succeeded because I have never so much as scratched a passenger — although in the early days I made a lot of them sick! I've always done everything to try and convince the public that airline flying is humdrum and safe.

Loganair have named a succession of Islander aircraft after Eric to mark his contribution to the Scottish Air Ambulance Service.

The DH-114 Heron — wheels down and welded.
*(Adrian Meredith Photography)*

Eric meets Eric — a portrait presented upon the occasion of his retirement from flying.

Eric shares a joke with Captain Ric Burroughs at the computerised flightdeck of British Airways' Boeing 757. The occasion was the fortieth anniversary of the Edinburgh-London service which Eric had inaugurated on 1st April 1949.
*(D.C. Thomson & Co. Ltd.)*

Michael Bishop, Chairman of British Midland Airways, Eric, and Scott Grier, Managing Director of Loganair, at a reception held in Inverness to mark the demonstration flight of Loganair's first jet aircraft, the BAe 146.
*(John Paul Photography, Inverness)*

Eric's last flight with Allied Airways was at the controls of Rapide G-ADAH, 'Pioneer' in 1939. Forty-nine years later they meet again at the Museum of Flight, East Fortune. G-ADAH is now located at Manchester Air & Space Museum.
*(Iain Hutchison)*

Ever game for new challenges, Eric has become a computer addict since retiring from flying.

A new generation of Starlings. Eric celebrates his 75th birthday with his grandchildren Allan, Katherine, Amy, Jonathan, Victoria and Hamish.

If I had a dodgy flight I regarded it as a failure and not something I would want to publicise.'

Eric could have continued working with the airline for a further five years on the ground. However he and Eleanor agreed that a full-time desk job was not for him and that his pension would be adequate for them both to live on. After three years of being standby air ambulance pilot, Eric had found many other activities demanding of his free time.

'B.E.A. offered me a ground job until I reached sixty-five, the compulsory retiring age. However Eleanor was strongly against it. She said that I would never be happy on the ground and, as I would be getting a good pilot's pension, we should enjoy it. And this we did. For five years we took full advantage of our concession travel. Eleanor was able to visit places I had spoken about but that she had never seen, and I was able to visit places I had visited but never seen. I have flown to Barcelona many times but had never been into the city until I retired. I have flown to Amsterdam countless times but, as yet, I have never been off the airport.

'Five years later Eleanor's heart was beginning to give trouble. However we had a very enjoyable holiday in Cyprus even though Eleanor was being bothered with angina attacks. On the way home we called in on our daughter, Elma, in Purley. There Eleanor had several bad angina attacks, but against the advice of the doctor, insisted on going home.

'On arrival home during the late afternoon, I immediately called our own doctor. He gave Eleanor a good examination and said that he would get her into hospital in the morning for a thorough check-over. 'That's not on', said Eleanor. 'I've to go to Brunei for the arrival of another grandchild.' 'You're joking', was the doctor's blunt reply. She died later that night.

'If I had taken that ground job, we would have missed five happy years. How wise was Eleanor and how I missed her.

'Once I had faced up to being an aged widower, I thought, 'The rest of my life starts now and I had better do something constructive about it.' It is so easy for old widowers to fade into the background. Eleanor was a first class cook so I had never done any myself. I decided that I must remedy this defect so that I could entertain visitors. To avoid loneliness one must attract visitors and, equally important, avoid adopting a habit of regularly having lunch at one of the local hotels. Cookery books are easy to follow these days and one has only to learn how to cook three good meals to be able to invite friends round three times without repeating oneself. I can honestly say that I never eat out when at home. Entertaining is so much more relaxed at home and it is amazing how female guests rush in to do the washing up. I doubt whether they would be so keen to do it for a widow. On one sad occasion my lady guest started her washing up by breaking one of my Edinburgh crystal glasses followed by a plate. However much I said that they were only objects and could be replaced, she felt dreadful and I could do nothing about it.

'A year or so later at a dinner, I was sitting next to a woman who did a radio programme, and I was telling her about a recipe that I had created. She liked the sound of it and said she would try it at home. If it was any good she would give it out on her programme. This she did and it boosted my morale no end.

'Elma's husband, my son-in-law, Ron, is in computing in a big way. He was always telling me that I should get a computer and in my sixty-ninth year I gave in and bought the cheapest computer that one could buy, a Sinclair ZX80. Much to my surprise I got hooked and have now progressed to the latest laptop computer. Once again my morale was boosted when, after two years computing, I sold a program I had written to an American magazine for $25.00. My computers also attract the local kids like bees to a honeypot. Computers are very robust and if you do the wrong thing they just don't work. I can therefore leave them to play computer games unsupervised knowing that they are learning to accept computers as a normal part of modern living. My home also lights up with their sounds of innocent merriment.

'It must be remembered that learning new things seems natural to a commercial pilot. All through one's career one is continually going back to school. Whenever there is a new aircraft or a new piece of equipment, back you go to the classroom. You have to be a permanent adolescent. Computer programming is the perfect hobby for someone retired because it is the supreme time-waster. They say you should always have a comfortable seat in front of your computer as a fifteen minute session so often becomes a two hour one. How right they are! It also keeps one's grey matter active. Brains and muscles have one thing in common. Stop using them and they atrophy.'

In recent years Eric has become a regular visitor to the U.S.A. During his career as a pilot he had never crossed the Atlantic despite numerous Coastal Command patrols which took him for countless hours above its grey waters and to within sight of the coast of Greenland. But that changed when he discovered some long lost cousins from the branch of the family that had changed their name back to 'Stirling' by deed poll in 1903.

This came about when Michael Stirling, a second cousin who was born near Sevenoaks in 1914, but now an American citizen living an individualistic lifestyle in New York's Bronx, sought Eric out during the course of some genealogical research. This in turn led to the discovery of more cousins living in Pleasant Valley, New York State; and at Hobe Sound, Florida. Eric also has a niece living to the south of San Francisco and a grand-niece in Calais, Maine. All of these recently-found relatives have now become firm friends resulting in frequent visits across the Atlantic by Eric.

One visit to Florida coincided with the night take-off of a Space Shuttle from Cape Canaveral and Eric was able to observe this from a safe distance. From Florida he flew on Eastern Airlines to San Francisco to visit his niece. 'The Captain had me up front for most of the flight. I was sitting just behind him as we were flying over Las Vegas. Seeing me trying to get a photo through the side window, he took out the auto-pilot and tilted the Tristar. The other three hundred-odd passengers never complained!'

'Ann Weller, who before marriage had been Ann Stirling, was researching her Scottish ancestory. When she discovered that I had family records she invited me over to their home in Houston. I remember saying to her that it was a bit of a gamble flying thousands of miles to visit complete strangers. We might have hated each other on sight. Fortunately we got on fine and I was very quickly one of the family. I've now visited them twice and they have also visited me in Largs.

'I once mentioned to them that the only things I had never flown in were an airship and a balloon. There was a Goodyear airship operating at Houston and they said they would organize a trip in it for me. Unfortunately it was away during my last visit so they arranged a trip in a hot air balloon instead. We had a very pleasant twelve mile flight over the flat Texan countryside finishing with the traditional champagne breakfast.'

The pilot in charge of the balloon was a white-haired white-bearded Neptune of the skies. Eric mentioned to him that at seventy-five this was his first balloon trip but Neptune didn't take the bait and reveal his own age.

However, four years later in February 1991, the Wellers rang Eric to say they had organised a trip in the local Goodyear Blimp. The Goodyear people said they didn't normally carry members of the public but in view of his distinguished career were prepared to carry him and six friends. So Eric took the next flight to Houston and on 22nd February 1991 enjoyed his first flight in an airship together with the Wellers and friends.

A Blimp is a non rigid airship. Eric was given the controls for part of the flight. He found a rudder controlled by foot pedals and elevators controlled by a large wheel operated by the pilot's right hand. After the flight Eric commented that flying at the dizzy speed of 30 m.p.h. he found it very unstable fore and aft and when the nose reared up he became concerned until he realised that he was flying nothing more than a powered, streamlined balloon which could hold that altitude indefinitely and didn't know the meaning of the word 'stall'.

Now the only thing left for him is a spaceship although he admits to having perhaps left that challenge a bit late. While in Houston Eric visited the Houston Astro Dome for his first rodeo. In the middle of the show everything suddenly stopped. President George Bush appeared on the large t.v. screen to announce the Gulf War ceasefire. The Bush connection continued with a visit by Eric to a Sunday morning service at the Houston Anglican Church. This is Bush's local church where he had escorted Margaret Thatcher during one of her visits.

Eric's North American visits have also taken him to Hollywood, Disneyland, to the N.A.S.A. Space Centre near Houston, Southfork Ranch in Texas where the American soap 'Dallas' is filmed, and to Calais on the border with Canada, home of a grand-niece. 'I have flown past and photographed Mount St Helens not long after it blew its top off. Twice I have had a flight delayed through bomb scares, once while the aircraft was searched at Los Angeles and the other time at Athens. On another occasion when I was attending a Croydon Airport Society meeting at the hotel beside the old airport, we all had to get out in a hurry because someone phoned to say a bomb was about to go off. Fortunately it was a false alarm. I think I can safely say that I haven't let the grass grow under my feet since I retired.'

Eric's travels have also taken him frequently to Brunei where Zoe and husband Tony, a Major in the Royal Fusiliers, were posted for nine years. Zoe had met Tony Hampton in Berlin, where she had been teaching the children of Forces' personnel, and where he was then stationed as a Captain. During one visit to Brunei, Eric joined a Royal Brunei Airlines flight at Singapore for the last stage of his journey and he viewed it with some trepidation when he discovered that the only other passengers were a pop group who had been invited to Brunei to give a series of concerts. The performers were a cosmopolitan mix of Chinese, Malays and Indians and their outlandish dress was only exceeded in its excesses by their bizarre hair styles. However once the flight had got underway Eric discovered that the pop group consisted of an extremely nice bunch of youngsters who were most interested in his reasons for visiting Brunei. The flight was both amusing and pleasant with Eric concluding that his initial doubts were the result of the pop group being obliged to assume the roles of actors playing their parts. This disguised the very real people underneath.

Eric is an accomplished speaker, especially on subjects close to his heart, like aviation. He is therefore frequently called upon by various organisations, such as the Largs Branch of the Royal Air Force Association of which he is President, to address gatherings. But his celebrity status has extended beyond this sphere, particularly as anniversaries of aviation mile-stones take on an ever growing significance.

On 23rd November 1977 he was on board a British Airways Hawker Siddeley 748 from Aberdeen to Sumburgh to re-enact the first air mail service between the two points forty years earlier. Joining him were Eric Gandar Dower, in his 83rd year and vowing to a dinner gathering after the event that 'this is positively my last public appearance', and his loyal companion Caroline Brunning. The flight took a mere 65 minutes (compared with three hours for the original flight) but some things never change — the programme on Shetland had to be curtailed by one hour to enable the return aircraft to depart early for Aberdeen as bad weather threatened to close Dyce Airport.

But Gandar Dower hadn't quite made his final exit from the public eye. On a return visit to Aberdeen he and Caroline Brunning, after nearly half a century of companionship, finally married on 7th April 1978 at the ages of 83 and 76 respectively. Eric was on hand to perform the duties of the Best Man.

Further trips to Shetland occurred in 1986 for the fiftieth anniversary of the first scheduled flight and again in 1987 for the fiftieth anniversary of the Air Mail service. On that occasion a 1930s Post Office van was on hand to deliver the mail. As Eric became increasingly called upon to attend these events he would introduce himself as 'a walking relic'. However as he strolled about the windswept tarmac at Sumburgh examining parked aircraft, he was eventually collared by an out-of-breath reporter who commented, 'That title may be all very well but I wish you would walk a little slower.'

While the Post Office had procurred a vintage Mail Van for the Shetland Air Mail anniversary, there was no vintage Allied Airways aircraft awaiting it on the tarmac. However the following year during the compilation of this book, Eric and I visited the Museum of Flight at East Fortune where we were able to view Allied Airways' de Havilland Rapide, 'Pioneer', in a storage hangar. Eric had last seen the aircraft forty-nine years previously when he had been at the aircraft's controls during his very last flight for the airline.

For the fiftieth anniversary of Stavanger Airport, Eric was invited by the Norwegians as a special guest since he had been recorded as the first international arrival at Sola following completion of his historic survey flight from Aberdeen in 1937. The Red Arrows had been invited to give a display and Eric was thrilled when they proved to be the highlight of the afternoon's celebration programme. After they had completed their spot they landed to refuel and then flew to Prestwick to give another display that same afternoon. How the North Sea had shrunk to an insignificant stretch of water compared with the vast expanse of sea confronting the Allied Airways adventurers on 22nd May 1937.

Eric was transported across the North Sea again two months later, joined by his son and daughter-in-law, Angus and Margaret. This time the hosts were Dan-Air, present incumbents of the Newcastle-Stavanger route using BAC1-11 jets, who were honoring Eric's first scheduled service on the route half a century before.

Loganair first honoured Eric in 1973 when they named Britten-Norman Islander G-AWNR after him and his name has subsequently been transferred to other Islander aircraft as the fleet has been updated. However Eric had never flown on Loganair until 1988 but that was rectified when he joined the guests on the demonstration flight, from Edinburgh to Inverness, of Loganair's first jet aircraft, a BAe-146.

To mark the fortieth anniversary of the inaugural B.E.A. service from Edinburgh to London on 1st April 1949, British Airways sent a car for Eric to join a celebration flight between the two capitals. The 189 passenger Boeing 757 jet, with its highly computerised instrumentation, contrasted starkly with the 22 passenger Douglas Dakota that Eric had captained for the original flight.

In reflecting on his long flying career this prompts Eric to note other contrasts.

'I am often asked for my observations on the enormity of the changes that I have seen since I started flying. I always pinpoint the most striking of these with, 'The size of aircraft and the size of pilots' salaries.'

'The old Dragon Moth that we used for flying round the islands could sit comfortably under a Boeing 747. When Scottish Airways was closed down and transfered to British European Airways, I was given a contract which stated that I would be eligible for transfer in the rank of Junior Captain at an annual salary of £900 and a Cost of Living Allowance of £76 per annum. During the years that followed, pilot's salaries increased substantially and I also found quick promotion to Senior Captain and then Flight Manager. Having enjoyed an index-linked pension for eighteen years, my monthly pension now compares favourably with the annual salary that I received upon joining B.E.A.

'I was recently watching a Jimmy McGregor programme featuring the island of Foula. Foula and Fair Isle are the only two islands that I regret never having visited.

'While watching the programme, I became very aware of the fact that the only thing I miss about flying is flying round the Scottish islands. Flying up and down the airways in a very controlled manner I am quite happy to forget. What I miss is Gordon Grant, the Controller at Kirkwall, who addressed the Captain by his first name on the first radio call. I miss splashing down on the sands at Barra ('Wind and tide permitting' as it said in the timetables) and Kitty Macpherson's delightful voice over the walkie talkie she used for communicating with us. I miss Dr Macleod of Lochmaddy and his wife, also a doctor. Dr Macleod would himself carry the patient into the aircraft and the patient would then have to wait while we were taken over to their Land Rover to be plied with coffee and sandwiches while he got all the latest gossip.

'I miss the warmth, affection and appreciation one received from the islanders and the hospitality received on night stops. The hotel proprietor who would say to his wife, 'Captain So-and-so is coming tomorrow. We must be sure to have kippers for his breakfast because he likes them so much'. I miss our arrival at the County Hotel, Stornoway, from where we would have a very early start. Stan Muir, the proprietor, would ask us, 'What time do you want a call tomorrow morning, Captain?' On being told one would be handed an alarm clock set at the right time. 'Do you want bacon and eggs as usual for breakfast?. . . Right I will see it is laid out for you in the kitchen'. I am an early riser so I usually ended up cooking my First Officer's breakfast as well. We carried our breakfast through into the dining room where a table was spread with fruit juice, etc. and we tucked into a good breakfast. Our passengers sat at another table eating sandwiches and drinking coffee poured from a thermos flask. Those were the days. I remember being taken round a curing shed in Stornoway. I noticed a salmon being smoked among the kippers. 'Oh, that's being smoked for one of your Captains', I was told. So different from the night stops we spent in the big cities of the U.K. and elsewhere where we were just a room number.'

# Epilogue

Eric Starling first shot to fame, or infamy, with his dramatic landing in the streets of Calais. The photograph showing his disabled aircraft, parked beside the monument of a hero from the local lifeboat, has sinced appeared in various publications over the years. When, in India during the Second World War, the same statue appeared in a film of the Normandy landings, Eric's colleagues thought he had been out in the sun too long when he announced, 'I've landed there!'

Not long after his emergency landing in Calais, Eric flew the first of many ambulance flights that he was undertake in the years ahead. It therefore transpires that his first meeting with the Calais statue perhaps held some symbolic premonition of the lives that Eric was to save with air ambulance missions during the coming years. I am indebted to Georges Fauquet for the story behind the statue of 'l'homme de barre'.

'On 18th October 1791 a trawler from Dieppe, the 'Saint-Pierre' with a crew of thirty-four, was shipwrecked near the jetty during a fierce storm. Sailors from the Courgain (the maritime district of Calais) rowed to their rescue and saved twenty-one men in five attempts. But after the fifth mission, and with the tempest growing worse, they were so exhausted that they couldn't move their limbs any more and were forced to give up.

'At this point four other young sailors decided to attempt to save the men who were still stranded on the wreck. They reached it and took the men aboard, but, as they were rowing back to the harbour, their craft was capsized. All the men in the sea clung to oars and anything else that was afloat. Then, Gavet, observing that one of the fishermen, one of the rescuers and himself were all clinging to the same oar, and knowing that his fellow rescuer had a wife and children, decided to give his companion and the fisherman a better chance of survival by voluntarily giving up his place on the oar to try and swim ashore. Another rescuer, Marichal, took similar action.

'Both were drowned and their bodies were found a couple of days later a few miles from Calais. The French government, 'The Convention', in want of heroes, profited by the incident to honour them. A monument was erected at the foot of the rampart of the Courgain. In 1899 a new monument, 'The Paradise', was unveiled and this is the one against which Eric Starling's plane landed so abruptly!

'During World War Two the Germans respected the monument. After the War it was moved and now stands in the new maritime district, facing the harbour entrance. The former Courgain was totally destroyed during the Battle of Calais in May 1940. The 'Homme de Barre' honours the self-sacrifice of Gavet and Marichal and depicts a Calaisian sailor throwing a rope to any person at peril in the sea.'

Of Eric's introduction to the Homme de Barre, Georges adds, 'I don't remember the accident, being too young at that time, but the following year (1934) my parents came to open a restaurant in the rue du Hâvre, a street leading to the monument, and people still talked about this event. A great aunt of mine, now dead, was an eye-witness to the incident as she was running a restaurant a few yards from the monument.'

# Milestones in the Sky

Captain Starling's flying career at a glance.

| | |
|---|---|
| 28 December 1911 | Eric Starling born at 70 Boundary Road, London |
| 11 August 1931 | First flight — with the Redwing Aircraft Company in Redwing G-AAUO |
| 8 October 1931 | First Solo Flight |
| 1 April 1933 | Aerial surveillance of The Boat Race |
| 16 May 1933 | Street landing in Calais at 1.00 am |
| 29 March 1934 | The Flying Fair opens at Southend |
| 19 June 1934 | The Flying Fair goes broke at Exmouth |
| 28 July 1934 | Opening of Dyce Airport, Aberdeen |
| 11 September 1934 | Aberdeen Airways inaugurate Aberdeen-Renfrew service |
| 27 May 1935 | Aberdeen Airways inaugurate Aberdeen-Stromness service |
| 4 June 1935 | Aberdeen Airways inaugurate Aberdeen-Edinburgh route |
| 11 June 1935 | Aberdeen Airways inaugurate Stromness-Thurso service |
| 2 February 1936 | First ambulance flight. South Ronaldsay to Stromness |
| 2 June 1936 | First scheduled service to Shetland |
| 5 June 1936 | Opening of Perth Aerodrome |
| 22 May 1937 | Aberdeen-Stavanger survey flight |
| 12 July 1937 | Allied Airways Newcastle-Stavanger inaugural service |
| 23 November 1937 | First Shetland Air Mail service |
| 2 May 1938 | 100th North Sea crossing |
| 19 September 1938 | 166th and final North Sea crossing |
| 14 January 1939 | Eric and Eleanor Castel marry at St Machar's Cathedral, Aberdeen. |
| 16 January 1939 | Commenced duties with Brooklands Aviation at Sywell |
| 3 September 1939 | War declared with Germany |
| December 1939 | Moved to Blackpool with No. 2 School of General Reconnaisance |
| 18 December 1939 | Angus Starling born at Blackpool |
| 22 November 1940 | Joined No. 221 General Reconnaisance Squadron at Bircham Newton |
| 2 May 1941 | No. 221 Squadron moves to Limavady and becomes operational |
| 1 October 1941 | No. 221 Squadron moves to Reykjavik |
| 11 October 1941 | Successful attack on U-boat |
| 3 January 1942 | Eric leaves Docking for Egypt |
| 23 June 1942 | Queen Mary sails from Suez |
| 3 August 1942 | Nieuw Zeeland docks at Stranraer |
| 22 August 1942 | Limavady again, this time with No.7 Operational Training Unit |
| 28 May 1943 | Officer Commanding No. 304 Ferry Training Unit, Port Ellen, Islay |
| December 1943 | Formation of No. 292 Air Sea Rescue Squadron |
| 23 March 1944 | Departure Thorney Island for India |
| 29 April 1944 | Arrive Jessore |
| 10 May 1944 | Zoe Starling born at Castle Douglas |
| 8 February 1945 | No. 292 Squadron moves to Agartala |

| | | |
|---|---|---|
| 30 March | 1945 | 5,000 hours flying recorded during flight from Akyab to Cox's Bazar |
| 4 May | 1945 | Landing in Rangoon Harbour |
| 1 August | 1945 | Last flight in the services — in a Supermarine Walrus |
| 2 January | 1946 | First flight for Scottish Airways |
| October | 1946 | Move from Inverness to Renfrew |
| 1 February | 1947 | Scottish Airways taken over by B.E.A. |
| 28 March | 1947 | Elma Starling born at Largs |
| 1 July | 1947 | Appointed Flight Captain by B.E.A. |
| 23 July | 1947 | Record breaking flight London-Glasgow with Viking 'Varlet' |
| 16 February | 1948 | Inverness-Stornoway Air Mail inaugurated |
| 1 April | 1949 | Launch of Edinburgh-London service |
| 1 July | 1950 | Renfrew-Perth service started |
| 15 May | 1951 | Australian Prime Minister flown Renfrew to Campbeltown |
| 4 September | 1954 | Campbeltown flight marks 20th anniversary of Aberdeen Airways exploratory visit |
| 31 May | 1955 | Service to mark 21st anniversary of first internal Air Mail, Inverness-Kirkwall |
| 7 July | 1959 | Princess Alexandra flown London to Milan |
| 24 May | 1962 | Official Royal Visit of Princess Alexandra to Stockholm |
| 26 July | 1962 | Prince Charles flown Inverness-Renfrew |
| 25 August | 1962 | Princess Margaret and Lord Snowdon flown Edinburgh-Aberdeen |
| 2 May | 1966 | First passenger flight to Glasgow Airport |
| 8 October | 1971 | Flight to mark 40th anniversary of first solo |
| 6 December | 1971 | Last flight, an air ambulance from Islay |
| 23 November | 1977 | 40th anniversary of Shetland Air Mail |
| 7 April | 1978 | Eric Gandar Dower and Caroline Brunning marry in Aberdeen |
| 2 June | 1986 | 50th anniversary of first scheduled service to Shetland |
| 22 May | 1987 | 50th anniversary of Sola Airport, Stavanger |
| 12 July | 1987 | 50th anniversary of first North Sea scheduled air service, Newcastle-Stavanger |
| October | 1987 | Flight by hot air balloon, Texas |
| 23 November | 1987 | 50th anniversary of Shetland Air Mail |
| 25 July | 1988 | Loganair BAe-146 demonstration flight |
| 1 April | 1989 | 40th anniversary of Edinburgh-London scheduled service |
| 22 February | 1991 | Goodyear Airship flight |
| 28 December | 1991 | Eric celebrates his eightieth birthday |

# Goggles to Gold Braid

During a flying career spanning over forty years Eric Starling has flown forty-five different types of aircraft. These range from open cockpit biplanes to multi-engined airliners. The roles for which they were designed include pilot training, joy-riding, aerobatics, surveillance, combat and passenger transport. Here are Eric's aircraft.

**Redwing Aircraft Co.**
Robinson Redwing
De Havilland DH60 Gipsy Moth
De Havilland DH80 Puss Moth
Southern Martlet
Avro Avian

**The Flying Fair**
Robinson Redwing
Klemm Swallow

**Aberdeen Airways/Allied Airways and associate companies**
De Havilland DH83 Fox Moth
De Havilland DH84 Dragon
De Havilland DH80A Puss Moth
Short Scion
Klemm Swallow
Klemm Eagle
Blackburn Bluebird
Blackburn B2 Trainer (Reserve Air Force Officers)
De Havilland DH89 Dragon Rapide
BAC Drone
Avro 504K
De Havilland DH60 Gypsy Moth
De Havilland DH86B Express
General Aircraft ST4 Monospar (demonstration)
Junkers Ju-52 (DNL Norwegian Airlines)
De Havilland DH82 Tiger Moth (RAFO and demonstration)
Witney Straight Special (demonstration)
De Havilland DH87B Hornet Moth (demonstration)
Praga Baby (demonstration)
Dart Kitten (demonstration)
Hawker Hart (instruction course)
Miles Hawk (demonstration)

**Brooklands Aviation**
Avro Anson

**No. 221 General Reconnaisance Squadron**
**(Bircham Newton, Limavady, Iceland, Western Desert)**
Vickers-Armstrong Wellington
De Havilland DH84 Dragon
Miles Magister

**No.7 Operational Training Unit, Limavady**
Avro Anson Lysander
Vickers-Armstrong Wellington
Miles Martinet

**No.304 Ferry Training Unit, Islay**
Beaufighter
Proctor

**Air Sea Rescue & No. 292 Squadron, Thorney Island**
Vickers-Armstrong Wellington
Vickers-Armstrong Warwick
Avro Anson
Miles Martinet
De Havilland Tiger Moth

**No. 292 Squadron, India**
Vickers-Armstrong Warwick
De Havilland Tiger Moth
Hawker Hurricane
Beechcraft Expeditor
Supermarine Sea Otter
Consolidated Liberator
Harvard
L5 Sentinel
Supermarine Walrus

**Scottish Airways**
De Havilland DH89 Dragon Rapide
Douglas DC-3 Dakota *
Junkers Ju-52 (Jupiter) *
*These were aircraft operating in B.E.A. livery in service pending the imminent formal start-up of the new Corporation.

**British European Airways Corporation**
De Havilland DH89 Dragon Rapide
Douglas DC-3 Dakota
Junkers Ju-52 (Jupiter)
De Havilland DH84 Dragon
Vickers Viking
De Havilland DH114 Heron
Vickers Viscount
Handley Page Herald

# Acknowledgements

In addition to acknowledging Eric's own considerable input to putting his story on to paper, I would like particularly thank Captain Tony Spooner DSO DFC for taking the trouble to prepare a lengthy narrative recalling the times he and Eric spent together at Brooklands, in 221 Squadron and during their later brief reunion in Malta.

I would also like to thank John Atkie and Bob Majors of the Museum of Flight, East Fortune, for making possible a trip down memory lane when we went to view Allied Airways' Rapide 'Pioneer'. And by way of a contrast I would like to thank Scott Grier, Managing Director of Loganair, for arranging that Eric join the demonstration flight of their first BAe 146 jet aircraft shortly after its delivery.

For responding to my enquiries I would like to thank David Beaty; Alan Cunningham, Chief Librarian, Western Isles Libraries; Iain Crosbie; Georges Fauquet, Société des Amis du Vieux Calais, Calais; F.R. Green; Jan-Petter Helgesen, Hafrsfjord, Norway; Wing Commander Jack H Hoskins; Olöf Magnùsdottir, Icelandair, London; Commodore Donald Murdo MacLean, former Master of the Queen Elizabeth; Group Captain A. Montagu-Smith DL RAF (RET.); Wing Commander Geoffrey H Porter RAF (RET.); Tor H. Strand, Airport Manager, Stavanger; and finally, Group Captain Tim Vickers.

For reading the manuscripts and providing useful suggestions I wish to thank Grahame Ross. For additional proof-reading I am also indebted to my mother, Patricia Hutchison. And lastly, but not least, thanks to Chuti, Panee and Narin for enduring the clack of typewriter keys into the wee sma' oors on many a night.

I am extremely grateful to Dugald Cameron who painted the picture from which the cover is reproduced. It depicts Aberdeen Airways' Dragon Rapide G-ADDE, *The Aberdonian* and Dragon G-ACAN, *The Starling* over the Old Man of Hoy in 1936. Dugald Cameron is Director of the Glasgow School of Art and author of 'Glasgow's Airport'.

Every effort has been made to trace the originators of photographic material. Apologies are made to anyone whose copyright may have been infringed or who may not have been given the appropriate credits.

Unless otherwise stated photographs are supplied courtesy of Eric Starling.

# Bibliography

**Air Road to the Isles.** Captain E.E. Fresson
Published by David Rendel 1967

**Flying against the Elements.** Peter V Clegg
Published by Peter V Clegg 1987

**A Flying Start to the Day.** Peter V Clegg
Published by Peter V Clegg 1986

**In Full Flight.** Captain A Spooner
Published by Macdonald & Co. Ltd., London, 1965

**Aviation in Scotland.** Gillies & Wood
Published by the Glasgow Branch of the Royal Aeronautical Society, 1966

**Scotland Scanned.** (Fifth Edition) Paul R W Wiggins & Alan J R Reid
Published by the Central Scottish Aviation Group, Edinburgh 1982

**Rebels and Reformers of the Airways.** R.E.G. Davies
Smithsonian Institution Press, Washington D.C. 1987

**The Story of Loganair.** Iain Hutchison
Western Isles Publishing Co Ltd, Stornoway 1987

**Kampen om Sola.** Jan-Petter Helgesen
Dreyer Bok, Stavanger 1987

**Croydon to Concorde.** Captain R E Gillman
John Murray 1979

**The Story of Aberdeen Airport.** James D Ferguson
Scottish Airports, Glasgow 1984

**Queens' Company.** Commodore Donald MacLean D.S.C., R.D, R.N.R.
Hutchinson & Co Ltd, London 1965